Archimedes

1996

Archimedes

NEW STUDIES IN THE HISTORY AND PHILOSOPHY OF SCIENCE AND TECHNOLOGY

VOLUME 1996

Archimedes, which will initially appear once a year, has three fundamental goals: to further the integration of the histories of science and technology with one another; to investigate the technical, social and practical histories of specific developments in science and technology; and finally, where possible and desirable, to bring the histories of science and technology into closer contact with the philosophy of science. To these ends, each volume will have its own theme and title and will be planned by one or more members of the Advisory Board in consultation with the editor. Although the volumes have specific themes, the series itself will not be limited to one or even to a few particular areas. Its subjects include any of the sciences, ranging from biology through physics, all aspects of technology, broadly construed, as well as historically-engaged philosophy of science or technology. Taken as a whole, *Archimedes* will be of interest to historians, philosophers, and scientists, as well as to those in business and industry who seek to understand how science and industry have come to be so strongly linked.

Archimedes

1996
New Studies in the History and Philosophy of Science and Technology

Scientific Credibility and Technical Standards in 19th and early 20th century Germany and Britain

edited by

JED Z. BUCHWALD

MIT and *The Dibner Institute for the History of Science and Technology,
Cambridge, MA, USA*

KLUWER ACADEMIC PUBLISHERS
DORDRECHT / BOSTON / LONDON

A C.I.P. Catalogue record for this book is available from the Library of Congress.

ISBN 0-7923-4762-5
ISSN 1385-0180

Published by Kluwer Academic Publishers,
P.O. Box 17, 3300 AA Dordrecht, The Netherlands.

Sold and distributed in the U.S.A. and Canada
by Kluwer Academic Publishers,
101 Philip Drive, Norwell, MA 02061, U.S.A.

In all other countries, sold and distributed
by Kluwer Academic Publishers,
P.O. Box 322, 3300 AH Dordrecht, The Netherlands.

Printed on acid-free paper

*Also published in 1996 in hardbound edition
in the series Archimedes, Volume 1996.*

Printed in Great Britain

TABLE OF CONTENTS

INTRODUCTION

ARCHIMEDES, which will initially appear once a year, has three fundamental goals: to further the integration of the histories of science and technology with one another; to investigate the technical, social and practical histories of specific developments in science and technology; and finally, where possible and desirable, to bring the histories of science and technology into closer contact with the philosophy of science. To these ends, each volume will have its own theme and title and will be planned by one or more members of the Advisory Board in consultation with the editor.

This first volume exemplifies two of our three goals, for the articles in it explicitly and intentionally cross boundaries between science and technology, and they also illuminate one another. The first three contributions concern optics and industry in 19th century Germany; the fourth concerns electric standards in Germany during the same period; the last essay in the volume examines a curious development in the early history of wireless signaling that took place in England, and that has much to say about the establishment and enforcement of standard methods in a rapidly-developing technology that emerged out of a scientific effect.

German optical products reigned supreme at the end of the 19th century, by which time the Zeiss name had become nearly synonymous with scientifically-designed and superbly crafted devices. The rise to preeminence of the Zeiss Werke in the 1870s was itself preceded by the remarkable reputation of German glass and lenses that were first produced at Joseph von Fraunhofer's Optical Institute in the 1820s. These developments are connected to one another by more than their having occurred in German-speaking lands, for in both cases successful industrialists based assertions concerning the particular superiority of their products on science. Towards the end of the century this binding between science and industry had become exceptionally tight, as the Zeiss Werke sought to capitalize on scientific effects and interests in order to produce extraordinary and unusual instruments that displayed the power and precision of the Werke's design and manufacturing capabilities.

Myles Jackson's Fraunhofer and Stuart Feffer's Ernst Abbe of the Zeiss Werke, though very different from one another in many ways, nevertheless share one particularly striking trait: both of them relied on scientific *arcana* to make claims for the goodness or aptness of their wares. Few had done so before them, but what makes the stories told here especially compelling is rather the character of their claims than that they were made at all. In Jackson's account we find Fraunhofer's line spectra, which Fraunhofer had first extracted with great skill from sunlight, and about which he alone had, at first, special knowledge, becoming effective tools for selling his lenses. Ernst Abbe, according to Feffer, turned a scientifically-problematic account of wave optics into an advertisement for his microscopes. Where the one marshaled an observational

J.Z. Buchwald (ed.), Scientific Credibility and Technical Standards in 19th and early 20th Century Germany and Britain, vii–ix.
© *1996 Kluwer Academic Publishers. Printed in Great Britain.*

and (for a brief moment) proprietary scientific novelty for manufacturing and marketing ends, the other used difficult but impressive-sounding scientific theory to transmute an industrial necessity into a virtue for creative advertising. Neither Fraunhofer nor Abbe was being the least duplicitous in what each of them did; both sought, admired and used scientific credibility where they could, and in so doing turned it into a fungible commodity.

David Cahan provides a thoroughly contextual, highly detailed account of the development early in the 20th century of the ultramicroscope at the Zeiss Werke, a device that could image colloidal particles lying below the theoretical limits of microscopical resolution that had previously been calculated (in quite different ways) by both Abbe and Hermann von Helmholtz. Here we find the Zeiss Werke taking advantage of the opportunity to, in Cahan's words, "explore previously uninvestigated or incompletely investigated scientific and technical ideas and fields that might contain sufficient potential for technological change and industrial profit". As in the case of Feffer's Abbe, here too the concepts and tools of wave optics entered directly into industrial practice. There is however a signal difference between the two situations. In the case of Abbe's light microscope, Feffer remarks, "optical theory as *publicly* employed by Abbe had far more to do with selling microscopes than with building them". The ultramicroscope, on the other hand, directly and explicitly deployed scattering and diffraction effects to detect sub-visible particles. Whereas, one might say, scientific strictures and effects were not themselves directly central to the industrial practices embodied in the Zeiss microscopes produced under Abbe, they had certainly become so a generation later for constructing the ultramicroscope.

Olesko's piece concerns the complicated scientific, cultural, and industrial issues that arose in the setting of electric standards in Germany, and, in particular, with the setting of a value for the ohm. Perhaps the most striking of the many aspects of standards work that Olesko unveils is the arresting difference between German and British sensibilities. In Britain, Olesko notes, individual differences in measurements were easily smoothed over, and error analysis of any kind was strongly downplayed. In Germany, by contrast, individual differences were made a very great deal of indeed, and discussions of error were common and contentious. In Olesko's words, German "precision represented an agreement between individuals, not among them as a single collective result".

The last paper in this volume, by Sungook Hong, is set entirely in England and tells an extraordinary story which, despite the locus, connects strongly to many of the themes set out in the other contributions. In the early 1900s, we learn from Hong's account, an attempt by Marconi and his technical adviser, John Ambrose Fleming, to demonstrate in public the efficacy of his wireless system was nearly undone by Nevil Maskelyne. Reputations suffered, tempers flared, credibility was angrily challenged, and the Marconi system, with its basis in tuned transmission and reception, was called into question. Indeed, Hong shows that much more than pure technical superiority was at issue in this affair. In order for the Marconi system to realize in full its technological potential it had to colonize the entire technical terrain; it could not reign successfully in a world of multiple devices. Maskelyne's tricks brought this directly into the open (and questioned as well Marconi and Fleming's moral character). Maskelyne felt he had used "the only possible means of ascertaining fact which ought to be in the possession of the public". Fleming called the effort an attempt to "wreck the exhibition". Both were correct, because each envisioned a unique economic and technical world. Where Maskelyne demanded an open universe of competing devices, Marconi and Fleming required a closed one that would be utterly dominated by a single type of equipment.

Historical work over the last few decades has shown that technology cannot be characterized simply, or even usually, as applied science. The beliefs, the devices, and the natural objects that are created or discovered by scientists often play altogether minor roles in the construction of technologies. Taking this realization as a given, the essays in this first volume of ARCHIMEDES effectively argue that we must now seek to go beyond it; we must begin also to think carefully about the roles that science actually did play when it was explicitly deployed by technologists. We need to ponder the implications of these stories and others for the real, historically-grounded connections that exist between scientific entities, the devices that make use of them, and the technological world. It seems to me that we should look as well for a novel philosophy of science, one that probes the nature of scientific work by grappling forthrightly and deeply with how it comes about that this particular form of human activity manages with such fecund regularity to produce novel entities that are inevitably bound to novel artifacts. Perhaps we need rather a philosophy of the "techno-scientific" than of either science or of technology alone.

MYLES W. JACKSON

BUYING THE DARK LINES OF THE SOLAR SPECTRUM: JOSEPH VON FRAUNHOFER'S STANDARD FOR THE MANUFACTURE OF OPTICAL GLASS

Joseph von Fraunhofer is of interest to physicists and historians of science because of his 'discovery' of the dark lines dissecting the solar spectrum, which now bear his name, and his advances in the manufacture of optical lenses. Previously, scholars discussing Fraunhofer's merits have erroneously placed him at the forefront of a long tradition of spectroscopy,[1] or have portrayed him as the 'father of German optics' in attempting to argue for a Germanic hegemony in both the physical sciences and optical industry.[2]

In this paper I shall argue that Fraunhofer's work on the dark lines of the spectrum was neither the culmination of any theoretical outlook on the nature of light nor an early effort in spectroscopy. Rather, it represented the labor of a self-educated artisan who wished to perfect the construction of achromatic lenses for astronomical instruments such as telescopes and heliometers, as well as ordnance surveying instruments such as theodolites. Fraunhofer's early work – including an early and influential experiment (discussed below) involving six lamps and solar light – was not an attempt to vindicate or debunk either the wave or corpuscular theory of light, despite the fact that his experiments most likely date from late 1813 to early 1815, precisely the period when British and French physicists were embroiled in intense debates on the nature of light.[3] Although Fraunhofer was to side with the wave theorists after his work on diffraction gratings in the early 1820s, he was never able to explain the dark lines. More than anything else, Fraunhofer's work was based on, and bound to, his technical mastery of artisanal knowledge and skill.

Fraunhofer's knowledge of optics, reflected in both the experimental design and in the calculation of the refractive indices for his glass prism samples, was quite basic. Indeed, for such purposes one only requires knowledge of (and practical skill in handling) Snell's Law for refraction, as well as basic trigonometric relations.[4] Fraunhofer deployed the law and its implications in his everyday work with glass prisms.

Precisely because Fraunhofer's contribution to physical optics emerged out of his manual dexterity and artisanal skill and was not based on any theory of the nature of light, the British could not merely resort to optical theory in order to reconstruct those skills. Indeed, the personal, uncommunicated character of Fraunhofer's artisanal knowledge guaranteed his Optical Institute's hegemony. The first portion of my essay analyzes Fraunhofer's extremely precise method for calculating refractive and dispersive indices of achromatic glass. Before his work, the determination of dispersive indices of glass prisms was restricted to a rather imprecise procedure whereby the optician would take the mean of the refraction of the extreme red and violet rays. This two-ray method had been necessary since before Fraunhofer's glass manufacturing techniques, prisms

1

J.Z. Buchwald (ed.), Scientific Credibility and Technical Standards in 19th and early 20th Century Germany and Britain, 1–22.

had not been good enough to demarcate precise portions of the solar spectrum. In such prisms red blended into orange, orange into yellow, and so forth. Fraunhofer, as we shall see, was able to determine the refractive indices for a much more precise portion of the solar spectrum by using the solar dark lines as boundary markers. Only his prisms – the result of unparalleled skilled manipulations of optical glass manufacturing – could show these lines. Extremely pure ingredients, special stirring techniques to ensure homogeneity and a novel method for polishing lenses perfectly symmetrically yielded Fraunhofer's first line-displaying prisms, which were subsequently used as calibration devices in producing even more accurate prisms and lenses.

The second part of this paper analyzes how Fraunhofer rendered certain portions of his enterprise public, such as the publication of the method using the dark lines as a calibration technique for producing achromatic lenses, while keeping other portions private, such as the method and recipe for manufacturing those achromatic lenses. Fraunhofer's choice of what to disclose and what to keep secret was very clever. By convincing prospective clients (both entrepreneurial and scientific) that his method of calibration (based on the specrtal dark lines) was the most accurate for producing lenses, and by demonstrating with that particular method that his lenses were indeed of a superior quality, Fraunhofer immediately increased the market for his Institute's products. By neither publishing the technique nor the recipes for glass making, nor permitting access by experimental natural philosophers to his glass hut, Fraunhofer ensured that the ever-increasing market would be compelled to purchase only his Institute's lenses; he guaranteed the market's fidelity. This story, in part, deals with an invention of a technique for testing quality that, once widely accepted, gave Fraunhofer and his Optical Institute a significant lead time for perfecting glass-making techniques. It is the goal of the second part of the essay to illustrate that the dark lines traveled throughout Europe, particularly Britain, much more readily than the artisanal skills needed to manufacture achromatic lenses. In short, Fraunhofer's plan was extraordinarily successful.

FRAUNHOFER AND THE OPTICAL INSTITUTE AT BENEDIKTBEUERN

The Optical Institute at Benediktbeuern was housed in a secularized Benedictine cloister which Joseph von Utzschneider, Bavarian entrepreneur and former Privy Councillor, purchased from the Bavarian government in 1805.[5] The Institute's main purpose was the manufacture of achromatic telescopes and ordnance surveying (military inventory and supplies) instruments for the topographical mapping project of the French and Bavarian *Bureau topographique* ordered by Napoleon. Napoleon's armies of occupation needed accurate maps of his new ally. Furthermore, King Maximilian I's economic reforms for Bavaria were fueled by property taxes, which required more accurate maps of the territory, and therefore, astronomical and ordnance surveying instruments. Astronomers, ordnance surveyors and map makers, for whose instruments Fraunhofer manufactured achromatic lenses, were his first clients.

Because Fraunhofer was an integral member of a business, the Optical Institute, strict measures were taken to differentiate between public and private knowledge. On the one hand, Fraunhofer desired to increase the visibility of the Institute in order to increase sales. Hence, he needed to publicize aspects of his enterprise. This was accomplished by publishing lists of the the Optical Institute's products as well as by entertaining visiting experimental natural philosophers and demonstrating to them his method for the calibration of achromatic lenses. Fraunhofer also wished to obtain credit for his work from the scientific community. This he was able to do by publishing his essay on the experiments which enabled calibration in the prestigious journal of the Royal Bavarian Academy of Sciences. On the other hand, certain forms of his artisanal knowledge needed to be kept secret. Hence, Fraunhofer policed both written word and social space. He published neither the procedures nor recipes for producing achromatic lenses, nor did he ever permit experimental natural philosophers to witness the skills involved in lens production.

The reputation of Fraunhofer's legendary workmanship quickly spread throughout the German territories, as ordnance surveying projects were being conducted in Prussia, Saxony, Lower Saxony and Schleswig-Holstein during this period. Fraunhofer's workmanship was matched only by Utzschneider's entrepreneurial genius. Utzschneider published the Optical Institute's price lists for achromatic telescopes, theodolites, borda circles, transit instruments and optical lenses in journals affiliated with ordnance surveying and astronomy. These lists appeared, for example, in von Zach's *Monatliche Correspondenz zur Beförderung der Erd- und Himmelskunde*,[6] von Lindenau and Bohnenberger's *Zeitschrift für Astronomie und verwandte Wissenschaften*[7] and even in optical texts such as J.J. Prechtl's *Praktische Dioptrik*.[8] Astronomers, ordnance surveyors and map makers across Europe became familiar with the Optical Institute's products.

Experimental natural philosophers from Britain, France and the German territories would visit the Optical Institute to inquire whether Fraunhofer could manufacture lenses to their specifications. For example, Gauss traveled to Benediktbeuren in 1816 to place an order for astronomical instruments for his new observatory at the University of Göttingen.[9] Other astronomers followed suit. Olbers from Bremen, Schumacher from Copenhagen and Bessel from Königsberg all ordered their astonomical instruments from the Optical Institute. His reputation was also enhanced by the scores of articles published by Bessel, Olbers and Struve announcing the discovery of stars hitherto unseen, but now made visible by Fraunhofer's refractors. And finally, of course, Fraunhofer's reputation soared as he continued to publish his research on light and physical optics in prestigious journals.

Visitors to Benediktbeuern were not restricted to astronomers and experimental natural philosophers. King Maximilian I; his son, King Ludwig I; Freiherr von Montgelas, the leader of Bavaria's reforms in the first two decades of the nineteenth century and the King of Prussia all visited the secularized cloister to see the Optical Institute. The Czar of Russia had originally planned to pay a visit upon his return from the Congress of Vienna, but cancelled his trip at the last moment.[10] He did, however, place an order for achromatic telescopes for his Russian universities, including the renowned refractor at Dorpat.

FRAUNHOFER'S SIX LAMPS AND SOLAR LIGHT EXPERIMENTS

Fraunhofer's essay "*Bestimmung des Brechungs- und Farbenzerstreuungs-Vermögens verschiedener Glasarten in bezug auf die Vervollkommnung achromatischer Fernröhre*" ("Determination of the Refractive and Dispersive Indices for Differing Types of Glass in Relation to the Perfection of Achromatic Telescopes") written from 1814 to 1815 was the culmination of his experiments on perfecting the manufacture of achromatic lenses.[11] Fraunhofer's essay was certainly not an attempt to explain theoretically the nature of the solar dark lines (later to be called aborption or Fraunhofer lines) or the lamp lines (later to be called emission lines). The sole purpose of this essay was to publish his method for improving the construction of achromatic lenses for telescopes. Indeed, the essay commences by claiming that:

The calculation of an achromatic object glass, and generally that of every achromatic telescope, necessitates a precise knowledge of the ratio of the sines of incidence and refraction [Snell's Law], and of the ratio of various types of glass which are used in the construction of telescopes ... Experiments repeated during many years have led me to discover new methods of obtaining these ratios, and I have therefore obeyed the wishes of several scholars [astronomers and experimental natural philosophers] in publishing these experiments, in the order I made them, with the necessary modifications that the experiments themselves forced me to introduce.[12]

He concludes his piece by admitting:

In making the experiments of which I have spoken in this memoir, I have considered principally their relations to practical optics. My leisure did not permit me to make any others, or to extend them any farther. The path that I have taken in this memoir has furnished interesting results in physical optics, and it is therefore greatly hoped that skilful investigators of nature would condescend to give them some attention.[13]

Ever since Isaac Newton's work on the spectrum during the late seventeenth and early eighteenth centuries, experimental natural philosophers and opticians had attempted, in vain, to determine the dispersion and refraction of each colored ray. But since colors of the spectrum appeared to be continuous, no precise limits could be detected. Fraunhofer himself admitted in his essay that:

It would be of great importance to determine for every species of glass the dispersion of each separately colored ray. But since the different colors of the spectrum do not present any precise limits, the spectrum cannot be used for such.[14]

He, as many others before him, attempted to cirmcumvent this problem by focussing his attention on colored glasses and prisms filled with colored fluids with the hope of determining the refractive index of the glass sample for that color of light. It was hoped that these glasses and prisms would permit only homogeneous light to pass. Despite various attempts to produce such a glass or fluid, the emergent light never proved to be truly monochromatic; a mixture of spectral colors always resulted. Fraunhofer also attempted to use the colored

Figure 1. A modified theodolite used by Fraunhofer to determine the refractive indices of his glass prisms. I would like to thank the Deutsches Museum, Munich for their kind permission to reprint this photograph (Fraunhofer Ordner 75391).

flames produced by burning alcohol and sulfur. But these flames, too, produced a spectrum when viewed through a prism.[15] He did, however, notice during these investigations that the spectra produced by the alchohol and sulfur flames were marked by a clearly defined line in the orange region. This line, later discovered to be two lines very close to each other and called the sodium lines, proved to be crucial to Fraunhofer's subsequent research. These lines served as the basis for Fraunhofer's "Six Lamps Experiment," which was designed to isolate homogeneous light.

After giving up on colored glasses and liquid-filled prisms, Fraunhofer returned to using white light, the source of which was the sodium lamp. He wished to view the spectrum produced by a sodium lamp, refracted by a prism and viewed through the telescope located on a modified theodolite made by Georg Reichenbach. Theodolites are ordnance surveying instruments designed

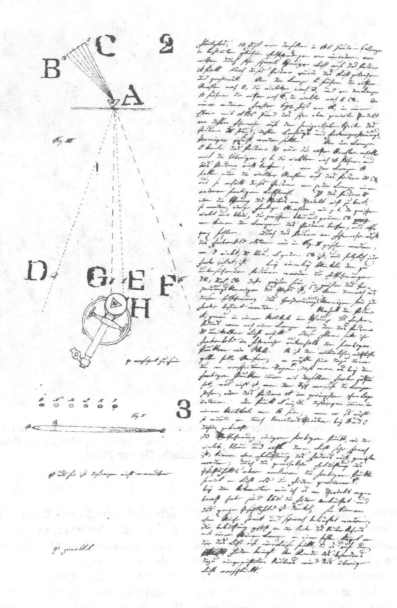

Figures 2 and 3. A page from Fraunhofer's manuscript of "Bestimmung des Brechungs-
und Farbzerstreuungsvermögens verschiedener Glasarten". The top portion (Figure 2)
illustrates his Six Lamps Experiment. Figure 3, at the bottom of the photo, depicts what
an observer looking through the telescope of the modified theodolite would see when
sodium lamps are used as the light source. I would like to thank Deutsches Museum,
Munich for their kind permission to reprint this photograph (Fraunhofer Ordner 33018)

to measure angles for maps. The modified theodolite could measure the angle of emergence from the prism for each colored ray (see Figure 1). Unfortunately, the rays of light falling onto the subject prism would not be parallel so that the angle of incidence would not be the same for each one, rendering the modified theodolite's measurement useless. In order to ensure that the rays striking the prism would be parallel, Fraunhofer substantially increased the distance between the lamp and the prism. But, he noted, although the rays now all had measurably the same angle of incidence, the increased distance resulted in some of the refracted rays missing the prism altogether. In order to ensure that 1) the incident rays remained parallel, 2) an entire spectrum would be generated and 3) the intensity of the light would be sufficient to be seen through minute slits at such a large distance, Fraunhofer used a total of six sodium lamps.

Fraunhofer placed the six lamps behind a shutter 1.5 Bavarian inches (slightly less than two English inches) high and .007 inches thick (see Figure 2). The shutter was pierced by six narrow slits slightly less than 1.5 inches high and .05 of an inch wide. The six lamps were placed .58 of an inch from each other directly behind the shutter, putting each lamp centrally behind one of the slits. The shutter itself was placed 13 Bavarian feet (or slightly more than 14 English feet) from prism A, which was made of flint glass and had an angle of approximately 40°. The light, having passed through the slits, was now refracted by the prism and decomposed into colors. The dispersed light then travelled through a second slit placed directly behind the prism, which accordingly blocked a portion of the emergent beam. Some of the rays were channelled to the site of a modified theodolite located in Fraunhofer's laboratory at the very great distance of 692 Bavarian feet (or 225 meters) from the six lamps (Figure 1).

The six-shutter mechanism controlled the angles at which light from each lamp struck the surface of prism A, thereby determining the locus of the corresponding spectrum. For example (Figure 2), from the lamp at C, red rays refracted to E and violet to D. From lamp B the red rays travelled towards F and the violet rays towards G. On the theodolite Fraunhofer placed a prism H whose index of refraction for the different colored rays was to be determined. He then adjusted the distances of the six-shutter mechanism from prism A, of A from the single shutter, and of the single shutter from prism H in such a manner that prism H received only red rays from lamp C and only violet rays from B. The intermittent lamps supplied the other colors of the spectrum.[16] The spectrum of rays passing through the small aperture below A and then through prism H will appear in the modified theodolite's telescope as depicted in Figure 3, where I is the violet, K the blue, L the green, and so on; each spectral color will appear at a unique locus. Fraunhofer ground down the angle of prism H until all the rays from the six lamps sensibly emerged from H at a single point (though, of course, each colored ray exits from that point at a unique angle with respect to the face normal there). The object lens of the theodolite's telescope was aimed at that point, thereby enabling Fraunhofer to

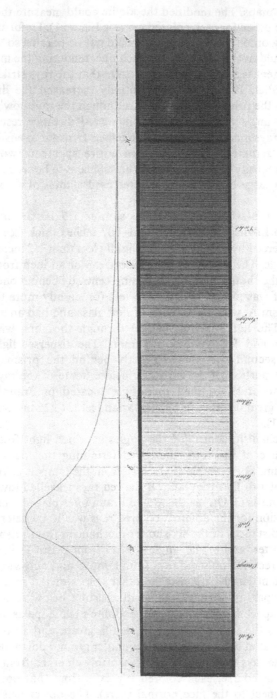

Figure 4. The solar spectrum with the dark (later to be called Fraunhofer) lines as depicted by Fraunhofer in 1814. I would like to thank the Deutsches Museum, Munich for their kind permission to reprint this photograph (Fraunhofer Ordner 43952).

see the entire spectrum and measure each colored ray's dispersion. The distances ON, NM, etc. (see Figure 3) increased with the dispersive power under these conditions. Since these distances and the incident angle could be measured by the modified theodolite to the nearest one-hundredth of an arc-second, Fraunhofer could now determine the index of refraction for each colored ray for each type of refracting substance.

In order to see whether other sources of light produced the same sort of lines as the sodium lamps had, Fraunhofer decided to use the sun as his source. He placed his theodolite and prism in a darkened room with a window that was covered by a shutter. He cut a vertical slit in the window shutter, 15 arc-seconds wide and 36 arc-minutes high (approximately 0.6 mm wide by 80 mm high) with respect to the center of the theodolite allowing the solar rays to fall on a flint glass prism with an angle of 60° mounted on the theodolite 24 feet from the window. The prism was placed in front of the telescopic objective lens in such a manner as to ensure symmetric passage (because, as Newton had argued over a century earlier, that position minimizes the effect of errors in setting the incidence).[17] Fraunhofer remarked:

In looking at this spectrum for the bright [sodium] line, which I had discovered in a spectrum of artificial light, I discovered instead an infinite number of vertical lines of different thicknesses. These lines are darker than the rest of the spectrum, and some of them appear entirely black.[18] (Figure 4)

Fiddling with the window-shade aperture and varying the distance of the theodolite from the window did not obliterate the lines. Between B and H, Fraunhofer counted 574 dark lines. Because the lines persisted no matter how he re-arranged the distances, Fraunhofer became convinced that these lines were not an experimental artifact, but were an inherent property of solar light.

This leads us to Fraunhofer's second major contribution to experimental optics. He was clever enough to use those dark lines as a natural grid that demarcates minute portions of the spectrum. Refractive indices could now be obtained for an extraordinarily precise portion of the spectrum: "As the lines of the spectrum are seen with every refractive substance of uniform density, I have employed this circumstance for determining the index of refraction of any substance for each colored ray."[19] He then chose the most obvious (i.e. the thickest and clearest) lines for his determination of the refractive indices of a glass prism: B, C, D, E, F, G and H (see Figure 4). They could easily be aligned with the cross-hatchings on his theodolite. Fraunhofer would simply read off the angle from the instrument's vernier. He made five measurements for each line.

To compute the index Fraunhofer used a standard equation found in several eighteenth-century optics textbooks for prisms.[20]

If δ = the angle of the incident solar ray

ρ = the angle of the emergent ray

ψ = the angle of the prism

n = the index of refraction

then $n = \sqrt{(\sin\rho + \cos\psi \, \sin\delta)^2 + (\sin\psi \, \sin\delta)^2} \, / \sin\psi$ (See Appendix)

Fraunhofer devised his experiments such that the angle of the incident ray was equal to that of the emergent ray (i.e. symmetric passage for the D line). If μ (the angle of deviation) is the angle between the incident and the emergent rays, then, under these circumstances,

$n = \sin[(1/2) (\mu + \psi)]/\sin[(1/2) (\psi)]$.

The angle μ was measured by the theodolite, as well as the arcs BC, CD, DE, EF, FG and GH. If n_E is the refractive index for the ray E, then

$n_E = \sin[(1/2) (\mu+\psi+DE)]/ \sin[(1/2)\psi]$;
$n_F = \sin[(1/2) (\mu+\psi+DE+EF)]/\sin[(1/2)\psi]$;
$n_G = \sin[(1/2) (\mu+\psi+DE+EF+FG)]/\sin[(1/2)\psi]$ and so on.

Fraunhofer created scores of tables listing the refractive indices of the rays (each ray corresponding to a line) for different substances: flint glass, crown glass, oil of turpentine and water, to name just a few. He then created tables of indices for combinations of refracting media in order to determine the combination that would correct chromatic aberration for the red and violet rays of the spectrum.[21]

Fraunhofer had now provided opticians and experimental natural philosophers with a vastly more precise method for determining the refractive indices of glass samples than had ever before been attained. Previously, Fraunhofer himself had determined the relative dispersive and refractive indices of two kinds of glass by cementing them together, forming a single prism. If the two spectra produced by this compound prism appeared at the same place, without any reciprocal displacement, he concluded that their dispersive and refractive powers were the same and equal to the arithmetic average of the two extreme rays: red and violet. After the discovery of the lines, however, he quickly realized that two pieces of glass, which appeared to have the same refrangibility when employing the early method of testing, could actually have slightly different powers, as revealed by the existence in the overlap region of two lines where there should be only one. Indeed, so sensitive was this new method that a difference in refracting power was found not only in different types of glass or between the same glass sample taken from different levels of the same melting pot or crucible, but even between pieces taken from the opposite ends (top and bottom) of the same piece of a glass blank. A high concentration of lead oxide in flint glass resulted in glass blanks being denser at the bottom of the crucible, since lead oxide is much denser than the other ingredients. Such a density gradient would destroy any attempt to produce achromatic lenses.

Fraunhofer and Pierre Louis Guinand now invented a method of stirring so efficient that, in a crucible containing over 400 pounds of flint glass, two specimens taken from the top and bottom had exactly the same refractive index. Overcoming heterogeneity, as we shall see, was not a problem for Fraunhofer, but it was a major one for Michael Faraday and for British attempts to 'reverse engineer' Fraunhofer's optical glass.

The dark lines of the solar spectrum, then, provided Fraunhofer with a tool for gauging the efficacy of achromatic lens production. If the refractive indices determined by aligning the dark lines with Reichenbach's theodolite indicated that the glass was not suitable for constructing an achromatic lens (i.e. if another lens could not correct the first lens' chromatic aberration efficiently enough), Fraunhofer and his assistants would alter the recipe by adding more or less lead oxide, or by altering the time of stirring or cooling, or by altering the ways in which the glass blanks were cut and ground into lenses. This portion of the process was trial-and-error, or an example of *"pröbeln"*, a Bavarian term similar to the German word *"proben,"* meaning to test, or try out. *Pröbeln* was a process that, inherently, could not be precisely specified. It is a type of knowledge that, by its nature, does not lend itself to replication. It is part and parcel of artisanal knowledge since it is only the skilled artisan who knows when the trial has indeed been a successful one.[22] Through trial-and-error technique, tested time and again by his solar-lines measurement technique, Fraunhofer produced the world's most coveted achromatic lenses.

FRAUNHOFER LINES IN BRITAIN

As I have argued elsewhere, Fraunhofer's calibration technique became the standard for optical lens production throughout Britain and the European Continent.[23] David Brewster wrote in 1825 that it was to Fraunhofer's essay on the six lamps experiment that one must turn for the "most accurate knowledge" in achromatic lens calibration.[24] An anonymous article in *The Edinburgh Journal of Science* defined an excellent prism as one that could clearly produce many Fraunhofer lines.[25] The British were now admitting that a Bavarian had surpassed their fellow countryman, Newton, in setting the standard for excellence in prism manufacture. While explaining Newton's analysis of the colors of the spectrum in his 1831 work entitled *A Treatise on Optics*, Brewster confessed that "no lines are seen across the spectrum . . . , and it is extremely difficult for the sharpest eye to point out the boundary of the different colours" in a replication of Newton's *experimentum crucis*. Brewster then proceeded to compare diagrammatically the lengths of the spectral colors as experimentally determined by Newton with Fraunhofer's results obtained by using a flint glass prism manufactured in the Optical Institute in Benedikt-beuern.[26]

Brewster was the first of the British experimental natural philosophers to recognize the importance of Fraunhofer's work to the discipline of optics. Indeed, he translated Fraunhofer's *"Bestimmung des Brechungs- und Farbenzer-streuungs-Vermögens verschiedener Glasarten"* of 1817 and published the translation in two parts (vols. 9 and 10 of 1823 and 1824) in *The Edinburgh Philosophical Journal*. During 1822 and 1823 Brewster was attempting to solve the problem of chromatic aberration in microscopes by using monochromatic lamps.[27] Hence, Fraunhofer's technique of using the spectral dark lines as a more precise system of calibration for determining the refractive indices aroused Brewster's interest.

John Herschel also played a major role in the technological transfer of the dark-line calibration scheme devised by Fraunhofer. During September of 1824, Herschel traveled to the Continent where he met Fraunhofer on the 19th September. During his visit Herschel hoped to obtain crucial information on the production of Fraunhoferian lenses for the Joint Committee of the Board of Longitude and the Royal Society for the Improvement of Glass for Optical Purposes. But Fraunhofer would not permit Herschel to witness the labor practices involved in optical glass manufacuring. He, from considerations of secrecy, reckoned that witnessing the labor practices would give Herschel and the British competitors too much insight. Fraunhofer did, however, demonstrate to Herschel how one could produce the dark lines in the spectrum and how those lines could be used as a system of calibration for achromatic lenses.[28] He could, therefore, on the one hand safeguard his Institute's market (and his income, since he received a percentage of the royalties generated from the instruments sold), while on the other hand promote his reputation as a *Naturforscher*, or investigator of nature, amongst his scientific colleagues.

We are told of Herschel's replication of the Fraunhofer lines in Charles Babbage's *Reflections on the Decline of Science in England*:

A striking illustration of the fact that an object is frequently not seen, *from not knowing how to see it*, rather than from any defect in the organ of vision, occurred to me some years since, when on a visit at Slough. Conversing with Mr Herschel on the dark lines seen in the solar spectrum by Fraunhofer, he inquired whether I had seen them; and on my replying in the negative, and expressing a desire to see them, he mentioned the extreme difficulty he had had, even with Fraunhofer's description in his hand and the long time which it had cost him in detecting them. My friend then added, 'I will prepare the apparatus, and put you in such a position that they shall be visible, and yet you shall look for them and not find them: after which, while you remain in the same position, I will instruct you *how to see them*, and you shall see them, and not merely wonder [why] you did not see them before, but you shall find it impossible to look at the spectrum without seeing them.'

On looking as I was directed, notwithstanding the previous warning, I did not see them; and after some time I inquired how they might be seen, when the prediction of Mr Herschel was completely fulfilled.[29]

Herschel's ability to replicate the Fraunhofer lines was crucial, as it rendered Fraunhofer's calibration technique credible. Herschel was, after all, Britain's leading experimental natural philosopher at the time, so his accepting the technique was tantamount to conceding that Bavarian lenses were indeed superior to English ones, as English prisms could not produce as many dark lines so clearly as Fraunhofer's. It should be noted that Herschel had observed spectral lines as early as 1821 and knew of the existence of a small number of solar dark lines.[30] A second reason why the British accepted the dark lines as representing a natural grid for calibrating lenses, rather than arguing that they were merely artifacts produced by inferior prisms, was the previous research which had been conducted by William Hyde Wollaston. Wollaston, who was a member of the aforementioned Joint Committee, had in 1802 published a short paper in *The Philosophical Transactions* entitled "A Method of Examining Refractive and Dispersive Powers by Prismatic Reflexions" detailing an experiment whereby he observed seven dark lines dissecting the solar spectrum. Hence, the dark lines were known previous to Fraunhofer's 'discovery.' Finally, because the dark lines could be incorporated into and explained by both the wave and corpuscular theories of light, neither side had any reason to question the phenomenon.

As James has argued, Brewster's and Herschel's works on absorption lines in the spectrum followed their own research interests in the 1820s.[31] But during the 1820s, Brewster raised the problem of accounting for absorption of light by the wave theory. Fraunhofer's lines were about to enter into theoretical technologies[32] which were absent in Bavaria. Brewster's work in early 1832 on nitrous acid gas (NO_2) revealed that it produces over a thousand dark lines in the spectrum of ordinary flames. He believed that this bore "strongly on the rival theories of light."[33] In his "Report on recent progress of optics" delivered at Oxford in June of 1832, Brewster claimed that the absorption must be a result of atoms which comprise the absorptive media. He could not see (or did not want to see) how to explain the absorption via the wave theory. Indeed, Brewster claimed that absorption could not, as those partial to wave optics insisted, be caused by molecules of a medium with complicated structures preventing the passage of light rays.[34] By 1833 Brewster publically attacked the wave theory for its inability to explain the spectral absorption lines.[35] After initially attempting to refute Brewster's claim, both G.B. Airy and William Whewell admitted that the wave theory could not satisfactorily explain the lines.

Herschel set out to meliorate the damage by evoking an analogy with sound. Since sound, which was generally believed to be propagated by waves, could experience absorption, light waves could, in theory, as well, if one could envisage a plausible analogy based on an acoustic mechanism.[36] By using tuning forks to construct a model for an absorbing system, for example, Herschel demonstrated that a system might vibrate and yet not send out waves; furthermore, and more to the point, this particular system did not permit other waves of the tuning forks' frequencies to pass.[37]

Fraunhofer was far removed from such issues. He had produced a technology adapted to the manufacture of achromatic lenses for Napoleon's ordnance surveying of Bavaria. He did not attach his lines to a pre-existing theory of light and never went back to explain these even after himself adopting a wave theory of light during the early 1820s with his work on diffraction gratings and interference. When this technology was transported to Britain, its direct association with ordance surveying was dropped altogether; it was instead brought into an existing framework of physical optics and theoretical accounts of light.

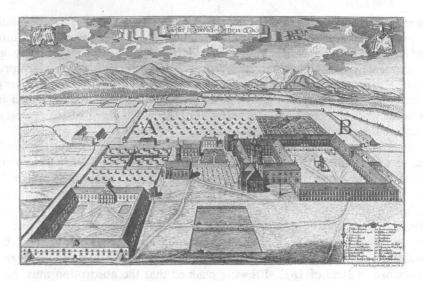

Figure 5. A woodcut of the Cloister Benediktbeuern from circa 1800, the site of Fraunhofer's Six Lamps and Solar Light Experiments as well as his manufacture of optical glass. The letter 'A' denotes the glass hut, whereas 'B' depicts Fraunhofer's laboratory. I would like to thank the Deutsches Museum, Munich, for their kind permission to reprint this photograph (Fraunhofer Ordner 30270).

FRAUNHOFER'S LENSES IN BRITAIN

Although the dark lines traveled in optical circles throughout Britain and the Continent, Fraunhofer's secrecy shrouding the manufacture of achromatic lenses did not. He restricted access to sensitive parts of his Optical Institute, preventing visitors from witnessing the skilled labor practices of glass production. By drawing upon a pre-exisiting Benedictine culture of secrecy that was built into the architecture of his laboratory, the Cloister of Benediktbeuern, Fraunhofer was able to establish a gradient of privacy in his scientific, technological and private enterprise.[38] The glass hut (letter A in Figure 5) was itself completely barred to visiting experimental natural philosophers such as Morelle de Serres, John Herschel and Carl Friedrich Gauss. These experimental natural philosophers, however, were encouraged to visit Fraunhofer's laboratory (letter B in Figure 5), where Fraunhofer could personally demonstrate to them his method of using the solar lines for calibrating achromatic lenses. Indeed, his laboratory was the most public place to visitors in the secularized cloister.

Fraunhofer's assistants originated in cultures where secrecy was paramount. Some of his assistants were Benedictine monks, whose vows of secrecy to God guaranteed that they would not divulge Fraunhofer's knowledge to anyone. Other assistants included skilled and unskilled laborers from the surrounding communities who closely guarded their guild secrets of glass making. The history of the practice of glass manufacturing in the forests of Bavaria and Bohemia is saturated with examples of secrecy. For example, glass makers often encoded recipe books with mysterious symbols, dechiperable only to those who studied under a particular glass master. Glass masters rarely disclosed their family name, for fear of being abducted by rival villages.[39] More importantly, Utzschneider controlled what Fraunhofer published. Hence, although Fraunhofer published his method for calibration of achromatic lenses using the dark lines of the solar spectrum, he never published the recipes of his crown and flint glass or an account of the actual skills and practices inherent to optical glass making. It should be noted that such skills generally cannot be replicated by simply reading a written account of them.[40] Even so, Fraunhofer would not have been permitted to publish any information that might offer potentially helpful insights to competitors.

The British response to Fraunhofer's ever-extending lead in optical lens production was the creation of the aforementioned Joint Royal Society and Board of Longitude Committee for the Improvements of Glass for Optical Purposes on 1 April 1824.[41]

The President of the Royal Society [Humphry Davy] having observed that the present state of the glass manufactured for optical purposes was extremely imperfect, and required some public interference: it was resolved to require the President and Council of the Royal Society to appoint a Committee to confer with the resident members respecting the best mode of conducting such experiments as they may think necessary at the expense of the Board.[42]

It was the explicit goal of this Committee to produce glass that would rival Fraunhofer's.[43] Members included such luminaries as Humphrey Davy, Thomas Young, John Herschel, William Hyde Wollaston and Michael Faraday.[44] Faraday was undoubtedly the most committed member of the Joint Committee. From 1825 to 1830 he was preoccupied with the chemistry of optical glass manufacturing.

As Britain's leading chemist during the period, Faraday, not surprisingly, attempted to 'reverse engineer' Fraunhofer's lenses and prisms by using chemistry, i.e. he attempted to work backwards from Fraunhofer's final products in order to obtain an experimental procedure that would securely and repeatedly yield that product. When Faraday received Fraunhofer's lenses and prisms, he immediately proceeded to determine their chemical composition. After ascertaining their ingredients in kind and percent, he concluded that he could indeed construct equally accurate lenses and prisms since the ingredients were standard and easily procured in Britain. But as time went on with no success in sight, Faraday began to realize that German flint glass was simply impossible to reproduce: "But be it remembered that it is not a mere analysis, or even the developement [sic] of philosophical reasoning that is required: it is the solution of difficulties, which, as is the case of Guinand and Fraunhofer, required many years of a practical life to effect ... ."[45] This quotation was from a paper which Faraday delivered as a series of lectures at the Royal Society of London in November and December 1829 and which was subsequently published in 1830 summarizing the results of his experiments on optical glass. Faraday continued by pointing out the possibility that "the knowledge they [Fraunhofer and Guinand] possessed was altogether practical and personal, a matter of minute experience, and not of a nature to be communicated."[46] As Faraday's own manipulative skills proved to be insufficient, he began to realize that the laborious process of lens and prism production could neither be obtained from the lenses and prisms themselves nor reconstructed merely from the knowledge of the ingredients and their percentages.

Since Faraday's attempt for the Joint Commitee to 'reverse engineer' Fraunhofer's lenses was proving to be unsuccessful, the British resorted to desperate measures, including bribery, offering Fraunhofer up to £25,000 for his secrets.[47] Fraunhofer did not take the British up on their offer, as he held to his written pledge to Utzschneider that any knowledge related to the production of achromatic glass was the Optical Institute's property, and therefore was not to be revealed to anyone other than the Institute's apprentices, and indeed then only sparingly. They then turned to Guinand's family members. In a Joint Committee meeting of 24 May 1827, it was

resolved that previous to incurring further expense in the manufacture of flint glass for optical purpose, the Board of Longitude be applied to, to authorize a personal application to the relatives of the late Mr Guinand, with a view to obtaining that portion of information which they may possess beyond ordinary glass manufacturers.[48]

After that attempt also failed, the Royal Society Committee decided to try to obtain information from the French optician Georges Bontemps. Bontemps had previously paid Pierre Louis Guinand's son, Henri, a significant sum of money for Guinand's optical glass making formula. The minutes of the 13 November 1828 meeting read: "resolved that Capt. Henry Kater [of the British Admiralty and Board of Longitude] be requested to make inquiries respecting the terms on which a knowledge can be obtained of the recent method employed of the late [sic] M. Bontemps[49] of making glass for optical purposes."[50] From 8 December 1828 until 6 January 1829 Kater, Peter Mark Roget, secretary to the Royal Society, and Bontemps entered into correspondence.[51] On 8 December 1828 Bontemps wrote to Kater discussing his and Mr. Chibaudeau's ability to manufacture optical glass for telescopes of large dimensions. According to Bontemps, his glass could rival the late Guinand's in terms of transparency and homogeneity.[52] Kater immediately responded by penning Royal Society President Davies Gilbert that he forwarded proposals from Bontemps "for the sale of the secret for making glass for optical purposes [. . .]. I beg to add that M. Bontemps has not received from me the slightest encouragement to expect the purchase either of the secret or the Dishes [the glass specimens which Kater purchased for examination by the Committee], but rather otherwise."[53] Bontemps' samples were, much to the outrage of the Committee, of grossly inferior quality, thereby proving that the required skills could not be so easily communicated in words alone. Dollond analyzed the glass and reported to Roget "the largest disk is extremely bad, and in my opinion very improper for the object-glass of a telescope. The smallest disk is not quite so imperfect as the largest, but in my opinion of little value."[54]

By late 1827 the English were beginning to realize that their attempt at 'reverse engineering', based on Faraday's chemical analysis, was a failure. Having next turned to Henri Guinand and Bontemps for assistance and having been disappointed here as well, they began to question whether or not the actual practices and labor of glass manufacturing might be so local as not to be accessible to them. Whether or not the skilled-artisans' practices of achromatic-lens production could be communicated at all, and if so how, became a central concern of the British experimental natural philosophers during the 1820s.

When tested with Fraunhofer's calibration technique, British lenses proved to be so clearly inferior to Bavarian ones that Fraunhofer's Optical Institute became the nearly exclusive purveyors of optical equipment to a vast market. The master craftsman's works were sold across the globe, including: Königsberg, Göttingen, Naples, Milan, London, Edinburgh, Paris, Marburg, Dorpat, Turin, Vienna, Regensburg, Berlin, Leipzig, Krakow, Lyon, Danzig, Ofen, Padua and Prague to name only a few.

CONCLUSION

This paper has offered an explanation of how the self-educated Bavarian skilled artisan, Joseph Fraunhofer, was able to develop a new calibration technique for the production of achromatic lenses. His work was not the culmination of optical theory, but represented a *Handwerker*'s trial-and-error method typified by skilled artisans. As he himself remarked in 1824, looking back at his technique:

There were not till the present time any fixed theoretic principles for the construction of achromatic object-glasses: and opticians were obliged, to a certain degree, to rely on chance, which made them polish a great number of glasses, and select those in which the faults could be most compensated. As the probability of this chance is much less in large glasses than in small ones, even those of the middle size would have been seldom perfect; and even with the best flint-glass the construction of large achromatic object-glasses would have been inpracticable. The more important causes which rendered this process necessary are as follows: The theory of achromatic object-glasses being as yet imperfect; the means formerly applied for ascertaining the powers of refraction and dispersion of colors in the different species of glass, which ought to rest on a firm basis, not being sufficiently established; and on account of the methods hitherto used for grinding and polishing the glasses not being calculated to follow the theory with that degree of exactness, as they ought, if a papable indistinctness should be avoided.[55]

Because Fraunhofer could convince his clients that his solar-line calibration technique was the most accurate method for testing achromatic lenses – and this apparently did not take much effort – and because he was not permitted to divulge his recipe or process for manufacturing achromatic glass, he was able to move his Optical Institute rapidly to market dominance. His conflicting interests (rendering certain portions of his enterprise public in order to achieve scientific recognition, while keeping other aspects private to ensure his Institute's hegemony) enabled Fraunhofer successfully to link science with manufacturing. As a result of Fraunhofer's success, Bavaria became, albeit ephemerally, the capital of achromatic glass production from 1815 until shortly after Fraunhofer's death in 1826. His method of optical glass manufacture and calibration technique remained the standard for lenses and prisms until the combined effort of Ernst Abbe and Otto Schott of the Carl Zeiss Works in the 1880s.

The University of Chicago, Department of History, the Fishbein Center for the History of Science and Medicine, and the Committee of the Conceptual Foundations of Science

NOTES

[1] See, for example: M.A. Sutton, "Spectroscopy and the Chemists: A Neglected Opportunity", *Ambix* xxiii (1976), 16–26; idem., *Spectroscopy and the Structure of Matter: A Study in the Development of Physical Chemistry* (D.Phil., Oxford, 1972) and W. McGucken, *Nineteenth-Century Spectroscopy: Development of the Understanding of Spectra 1802-1897* (Baltimore: The Johns Hopkins University Press, 1969), see particularly, 4–10. Frank James has correctly argued that such histories incorrectly describe the emergence of spectroscopy as a singular, continuous process of elaboration stretching back to Fraunhofer and extending forward from him to Gustav Kirchhoff and Robert Bunsen's establishment of the relationship between absorption and emission lines. Frank A.J.L. James, "The Creation of a Victorian Myth: the Historiography of Spectroscopy", *History of Science* xxiii (1985), pp. 1–24.

[2] A selected bibliography of Fraunhofer's hagiography includes: R. Bunsen and G. Kirchhoff, "Chemische Analyse durch Spectralbeobachtungen," in *Poggendorffs Annalen der Physik und Chemie* (1860), pp. 161–189; Hermann von Helmholtz, "Festbericht über die Gedenkfeier zur hundertjährigen Wiederkehr des Geburtstages Josef [sic] Fraunhofer's," in *Zeitschrift für Instrumentenkunde* VII (1887):114–128; Carl Max von Bauernfeind, *Gedächtnisrede auf Joseph von Fraunhofer zur Feier seines hundertsten Geburtstags* (Munich: Verlag der königlichen bayerischen Akademie der Wissenschaften, 1887); A. Seitz, *Joseph Fraunhofer und sein Optisches Institut* (Berlin: Julius Springer, 1926); *150. Todesjahr Joseph von Fraunhofers 1787-1826. Reden und Ansprachen*, ed. Fraunhofer-Gesellschaft zur Förderung der angewandten Forschung e.V. (Ingolstadt: Courier-Druckhaus, 1976); Sigmund Merz, *Das Leben und Wirken Fraunhofers*, (Landshut: Joseph Thomann'schen Buchhandlung, 1865); Ernst Abbe, "Gedächtnissrede auf Joseph Fraunhofer," in *Gesammelte Abhandlungen II, Wissenschaftliche Abhandlungen aus verschiedenen Gebieten Patentschriften Gedächtnisreden* (Hildesheim, Zürich and New York: Georg Olms, 1989), pp. 319–338 and the many works by Moritz von Rohr, including, most notably, his biography, *Joseph Fraunhofers Leben, Leistungen und Wirksamkeit* (Leipzig: Akademische Verlagsgesellschaft, 1929). Two more recent works have provided more nuanced accounts of Fraunhofer's life within a particular social and political context: Günther D. Roth, *Joseph von Fraunhofer. Handwerker-Forscher-Akademiemitglied*, 1787-1826 (Stuttgart: Wissenschaftliche Verlagsgesellschaft, 1976) and Hans-Peter Sang, *Joseph von Fraunhofer. Forscher. Erfinder. Unternehmer* (Munich: Dr. Peter Glas, 1987).

[3] See, for example, Jed Z. Buchwald, *The Rise of the Wave Theory of Light. Optical Theory and Experiment in the Early Nineteenth Century* (London and Chicago: The Univeristy of Chicago Press, 1989).

[4] The law was widely disseminated in eighteenth and early nineteenth-century texts. See, for example, P. Rogerii Josephi Boscovich, *Dissertationes Quinque ad Dioptricum*, (Vindobonae: Johannis Thomae, 1767) and Georg Simon Klügel, *Analytische Dioptrik in zwey Theilen. Der erste enthällt die allgemeine Theorie der optischen Werkzeuge: der zweyte die besondere Theorie und vortheilhafteste Einrichtung aller Gattungen von Fernröhren, Spiegelteleskopen, und Mikroskopen*, (Leipzig: Johann Friedrich Junius, 1778). Fraunhofer used the latter of these two texts extensively while manufacturing achromatic lenses.

[5] Deutsches Museum, Handschriften und Urkunden Sammlung, 5343 (Geschichte Fraunhofers und des Optischen Instituts).

[6] See, for example, XXVII (February 1813), pp. 197–9. Pater Ulrich Schiegg discussed the Optical Institute's merits during his report of the Bavarian triangulation project, "Astronomische Nachrichten aus Bayern," *ibid.*, XII (1805), pp. 357–63, here pp. 361–2.

[7] See, for example, II (1816), pp. 165–72.

[8] Johann Joseph Prechtl, *Praktische Dioptrik als vollständige und gemeinfaßliche Anleitung zur Verfertigung achromatischer Fernröhre. Nach den neuesten Verbesserungen und Hülfsmitteln und eignen Erfahrungen* (Vienna: J.G. Heubner, 1828), p. 178. Note that this list was published after Fraunhofer's death. But price lists were published in other optical texts from about 1818 onwards.

[9] Gauss ordered one meridian circle, two passage instruments, an equatoreal and two astronomical pendulum clocks from the Optical Institute. See, for example, Hans-Peter Sang (1987), *op. cit.*, note 2, p. 99.

[10] Staatsbibliothek Preußischer Kulturbesitz Berlin, Utzschneider Nachlaß Kasten 1, Briefe an Fraunhofer: 9, 15 and 16 July 1814.

[11] *Joseph von Fraunhofer's Gesammelte Schriften. Im Auftrage der Mathematisch-Physikalischen Classe der königlichen bayerischen Akademie der Wissenschaften*, E. Lommel (ed.), (Munich: Verlag der königlichen Akademie in Commission bei G. Franz, 1888), pp. 3–31, The article originally appeared in *Denkschriften der Königlichen Akademie der Wissenschaften zu München für die Jahre 1814 u. 1815*, Bd. V (not published until 1817). Fraunhofer sent this paper to his good friend, the mathematician, ordnance surveyor and member of the Royal Bavarian Academy of Sciences, Johann von Soldner. Inclusion into this periodical meant instant recognition as a *Naturwissenschaftler*, something that the working-class artisan desperately sought. Indeed, as a result of the merits of this essay, Fraunhofer became a corresponding member of the Academy in 1817. All references to this work will be cited Fraunhofer, followed by the page number corresponding to Lommel's edited collection.

[12] Fraunhofer, p. 3.

[13] Ibid., p. 27.

[14] Ibid.

[15] David Brewster also attempted unsuccessfully to isolate monochromatic light. David Brewster, "Description of a Monochromatic Lamp for Microscopial Purposes, & c. with the Remarks on the Absorption of the Prismatic Rays by Coloured Media," *Transactions of the Royal Society of Edinburgh* 9 (1823), 433–44 .

[16] Note that there can be an overlap of colors from the intermittent lamps, i.e. the lamp adjacent to B might supply prism H with a small portion of violet rays as well. However, since prisms map angle onto position, and since all of the incident rays strike prism H at the same angle (as a result of the large distance between the two prisms), all lamp light of the same wavelength will be mapped onto the same position as seen by the telescope of the theodolite.

[17] Note that the symmetric passage is for the D line.

[18] Fraunhofer, p. 10.

[19] Ibid., pp. 13–14.

[20] See note 4, Boscovich, p. 142.

[21] Chromatic aberration can be corrected in part by constructing a convex–concave lens combination made of flint and crown glass. This lens combination focuses the red and violet rays only into a point. It would take the combined effort of Otto Schott and Ernst Abbe in the 1880s to correct the red, violet and yellow rays.

[22] This part of Fraunhofer's scientific and technological enterprise is very analogous to cooking. Many master chefs argue that it is difficult to describe, particularly in the form of the written word, the attributes which inform the 'trained eye' that their creation is a success, such as its smell, taste or texture. Similarly, in glass making, knowing when and how long to stir, when the glass is cool enough to manipulate, etc. separates the master from the rookie apprentice. Indeed, sometimes apprentices never learn such skills. I would like to thank Julia Child for her helpful conversations on the necessary skills and practices for cooking and how such knowledge is communicated.

[23] Myles W. Jackson, "Artisanal Knowledge and Experimental Natural Philosophers: The British Response to Joseph Fraunhofer and the Bavarian Usurpation of Their Optical Empire," in *Studies in History and Philosophy of Science* 25 (1994), pp. 549–75, particularly 561–74.

[24] *Edinburgh Journal of Science* 2 (1825), 344–8, see p. 348.

[25] "Remarks on Dr Goring's Observation on the Use of Monochromatic Light with the Microscope", *Edinburgh Journal of Science* 5 (1831), pp. 153–8, p. 154.

[26] David Brewster. *A Treatise on Optics* (London: Longman, Rees, Orme, Brown and Green and John Taylor, 1831), p. 67.

[27] David Brewster, "Description of a Monochromatic Lamp", *op. cit.*, note 15. See also Frank A.J.L. James, "The Discovery of Line Spectra," *Ambix* 32 (1985), 53–70, here 59–60.

[28] Charles Babbage, *Reflections on the Decline of Science in England, and on Some of its Causes* (London: B. Fellowes, 1830), pp. 210–11 and *Edinburgh Journal of Science* 2 (1825), pp. 344–8.

[29] Babbage, *Reflections on the Decline, op. cit.* note 28, pp. 210–11 (italics in the original).

[30]   J.F.W. Herschel, "On a Remarkable Peculiarity in the Law of the Extraordinary Refraction of differently-coloured Rays exhibited by certain Varieties of Apophyllite", *Transactions of the Cambridge Philosophical Society* I (1822), 241–7, especially 244–5. This paper was read on 7 May 1821 at the Philosophical Society of Cambridge. Indeed, as will be discussed below, Herschel later attempted to explain those lines by using wave optics. See J.W.F. Herschel, "On the absorption of light by coloured media, viewed in connexion with the undulatory theory" in *Report of the British Association* (1833), 373–4. The full paper appeared in *The Philosophical Magazine* iii (1833), 401–12.

[31]   Frank A.J.L. James, "The Debate on the Nature of the Absorption of Light 1820–1835: A Core-Set Analysis" in *History of Science* 21 (1983), 335–68.

[32]   For the use of "theoretical technologies", see Andrew C. Warwick, "Cambridge Mathematics and Cavendish Physics: Cunningham, Campbell and Einstein's Relativity 1905–1911", *Studies in History and Philosophy of Science* 23 (1992), 635–56 and 24 (1993), 1–25.

[33]   James "The Debate" (1983), *c.f.* note 31, 340. See also David Brewster, "Observations of the lines of the solar spectrum, and on those produced by the Earth's atmosphere, and by the action of nitrous acid gas", in *Transactions of the Royal Society of Edinburgh*, vol. xii (1833), 519–30.

[34]   See James "The Debate" (1983) *c.f.* note 31, 340–41 for details.

[35]   David Brewster, "Observations on the absorption of specific rays, in reference to the undulatory theory of light" in *The Philosophical Magazine* ii (1833), 360–3, particularly 363.

[36]   Herschel, "On the absorption of light", *op. cit.*, note 30, 406.

[37]   J.W.F. Herschel, "Sound" in *Encyclopaeida metropolitana* ii (1830), 747–825.

[38]   Myles W. Jackson, "Illuminating the Opacity of Achromatic-Lens Production" forthcoming in *Architecture and Science*, edited by Peter Galison and Emily Thompson (Cambridge, MA and London: MIT Press, 1997).

[39]   Ibid.

[40]   Although Fraunhofer himself was convinced that one could construct his lenses and prisms only if one witnessed this artisanal labor, Utzschneider nevertheless guaranteed that the knowledge of optical glass maunfacture remained a secret by not permitting Fraunhofer to reveal the recipes. It is interesting to note that as Fraunhofer lay on his death bed, Utzschneider had his assistant take from the dying Fraunhofer an account of in the production of achromatic glass. Not surprisingly, Utzschneider could not replicate Fraunhoferian quality prisms and lenses from merely reading Fraunhofer's notes.

[41]   Cambridge University Library (Manuscripts Room) Royal Greenwich Observatory Archives (henceforth CUL.RGO) 14/8, folia 17–20. Frank A.J.L. James has argued that the Royal Society Subcommittee was created to bail out the financially doomed Board of Longitude. See his "Time, Tide and Michael Faraday", *History Today* (1991), 22–34, particularly 30–1 and his "The Military Context of Chemistry: the Case of Michael Faraday" in *Bulletin of the History of Chemistry*, 11 (1991), 36–40, particularly 37–8. I shall take another line, as the reader will see. I think that both James's point and mine need to be considered as joint reasons for the Subcommittee's creation. They certainly are not mutually exclusive.

[42]   CUL.RGO 14/8.20. This was an Act of Parliament.

[43]   Royal Society Committee Minutes Book (henceforth RS CMB).1.127 and The Royal Society Manuscripts (RS DM) Vol. III, Folia 26 and RS DM.III.22.

[44]   RS CMB.1. 96.

[45]   Michael Faraday, "On the Manufacture of Glass for Optical Purposes," in *Philosophical Transactions of the Royal Society of London* 120 (1830), 1–57, here, p. 4.

[46]   Ibid., p. 2.

[47]   Günther D. Roth. *Joseph von Fraunhofer. Handwerker-Forscher-Akademiemitglied 1787–1826* (Stuttgart: Wissenschaftliche Verlagsgesellschaft MbH, 1976), p. 120.

[48]   RS CMB.1. 158–9 and RS DM.III.41.

[49]   In the original maunscript, M. Guinand's name was written, then crossed out, and Bontemps's name was added. The 'late' referred to Guinand, not Bontemps.

[50]   RS CMB.1.211 and RS DM.III.53.

[51]   RS DM.III.63–70.

[52]   RS DM.III.63.

[53]   RS DM.III.62.

[54]   This was reported by Roget to Bontemps on 31 December 1828, RS DM.III.67.

[55]   Fraunhofer, "Ueber die Construktion des so eben vollendenten grossen Refraktors", *Joseph von Fraunhofers' Gesammelte Werke. op. cit.*, note 11, pp. 169–70; for an English translation of this essay, see *The Philosophical Magazine and Journal* lxvi (1825), pp. 41–7, here p. 43.

APPENDIX

Although there is no evidence that Fraunhofer actually derived an equation for calculating the refractive indices using prisms, one can very easily do so by using basic knowledge of optics, trigonometry and Snell's Law. Fraunhofer, as an optician, undoubtedly possessed such knowledge, and standard eighteenth- and nineteenth-century textbooks on optical theory and practice assumed such knowledge.

From Snell's Law applied to the entry (AB) and emergency (CB) faces of the prism ABC, one has (with reference to Figure 6):

$$\frac{\sin\delta}{\sin\beta} = \frac{\sin\rho}{\sin(\psi-\beta)}.$$

Simple mathematics yields:

$$\sin\beta = \frac{\sin\delta\,\sin\psi}{\sqrt{(\sin\delta\,\cos\psi + \sin\rho)^2 + (\sin\delta\,\sin\psi)^2}}.$$

Substituting back into Snell's Law, one obtains for the refractive index (n_{glass}) of the prism:

$$n_{(glass)} = \frac{\sqrt{(\sin\delta\,\cos\psi + \sin\rho)^2 + (\sin\delta\,\sin\psi)^2}}{\sin\psi}. \qquad (Eq.1)$$

In Fraunhofer's experiment the incident ray strikes at the same angle at which the emergent ray leaves (i.e., $\delta = \rho$). In this particular case of symmetrical passage through the prism we find (with reference to Figure 7):

$\mu \equiv$ the angle that the incident ray forms with the emergent, refracted ray.

1. $\alpha = (1/2)\psi$, from symmetrical passage. The deviation angle μ is the sum of the two opposite interior angles in the triangle aed, or $\mu = 2(\theta-\alpha)$, whence

2. $\theta = (1/2)(\mu+\psi)$. Since, at point a, θ is the angle of incidence, and α is the angle of refraction, then

3. $\sin\theta = n_{glass}\sin\alpha$, where n_{glass} is the index of refraction of the glass prism with respect to the air.

From 1, 2 and 3 we have:

$$n_{glass} = \frac{\sin\,[(1/2)(\mu+\psi)]}{\sin\,[(1/2)\psi]}. \qquad (Eq.2)$$

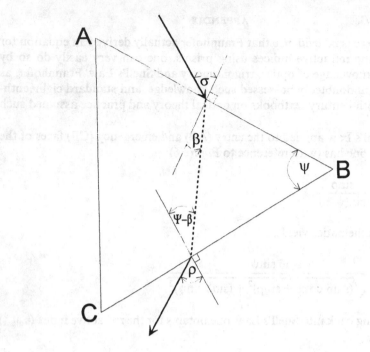

Figure 6.

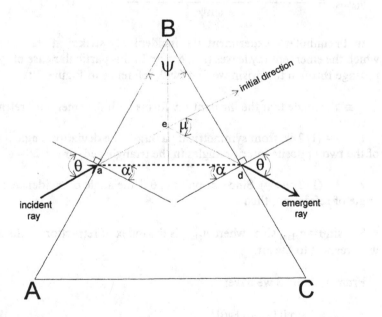

Figure 7.

ERNST ABBE, CARL ZEISS, AND THE TRANSFORMATION OF MICROSCOPICAL OPTICS

We have apparently come to a point where even the smallest improvements [in the performance of the compound microscope] can only be achieved with a disproportionately great expenditure of intellectual and mechanical work. Yet no one, so far as I can tell, has been able to give a single, specific reason why this should be so ...

> – *Hermann von Helmholtz in the* Annalen der Physik, *(1874)*

INTRODUCTION

For most of the nineteenth century, the high-powered compound microscope was a piece of technology that artisans could build, microscopists could use, but physicists could not understand. Like the steam engine earlier in the nineteenth century and several different electrical technologies (such as the telegraph cable) toward its middle, the microscope became as much an object of study as a tool of investigation or trade. In all three of these cases, those who would bring knowledge of academic experimental or theoretical physics to bear on practical problems had to face the – often well justified – resistance of more traditional practitioners whose practical and empirical knowledge of these technologies had served them well in the past.[1] The taming of the microscope by physicists required more than a theoretical account – a physical theory of microscopical optics had to convince microscopists that instruments designed according to that theory would be superior.

Microscopists themselves, not physicists, were the first to try to account for microscopical vision. They were immediately concerned to counter widespread distrust of the microscope by providing an analysis of microscopical optics and clear parameters for microscopical practice. Such a foundation, it was hoped, would settle controversies about microscopical techniques and equipment. Treatises on the microscope written by mid-19th century microscopists established a clear agenda of lingering questions to be answered about microscopical optics and problems to be solved by those who would supply new and improved microscopes. Ernst Abbe, physicist and partner in the optical firm of Carl Zeiss, delivered, producing a theoretical account of microscopical optics and new microscopes that exploited this theory with great empirical success. This paper traces the theory and practice of microscopical optics from artisanal craft to scientific practice, a transformation wrought in large part by scientist–entrepreneur Ernst Abbe.

During the last decade of the nineteenth century and the first decade of the twentieth, the group of researchers and optical designers that Abbe assembled at the Zeiss concern would broaden Abbe's techniques and apply them to a

23

J.Z. Buchwald (ed.), Scientific Credibility and Technical Standards in 19th and early 20th Century Germany and Britain, 23–66.

wide range of instruments. By the First World War, the Zeiss company, with subsidiaries and affiliated glassworks, had become one of Germany's largest industrial enterprises, dominating the world market in precision optical instrumentation. Its product line consisted of photographic apparatus, measuring instruments, astronomical telescopes, surveying instruments, range finders and binoculars – all designed using optical principles and materials directly traceable to Abbe's work on improving the microscope during the 1870s and 1880s.[2]

THE CRAFT OF MICROSCOPE MAKING

Early in the nineteenth century, the microscope was regarded more as a curiosity than as an instrument of natural philosophy. Plagued by imperfections, aberration, and lack of consensus about proper microtechnique, microscopic images often misled observers. Users of the microscope often found it difficult – if not impossible – to reproduce the findings of their colleagues.[3] One of the problems with these early instruments, shared by contemporary refracting telescopes, was chromatic aberration, colored fringes that appear around the edges of objects when viewed through the lens of a telescope. The search for ways to combat it occupied instrument makers and optical theoreticians alike for many years. The first achromatic telescope objectives had appeared by the end of the eighteenth century, consisting of one lens made from crown glass and a compensating lens made from flint, and opticians working on the perfection of achromatic telescope objectives also believed by corollary that their efforts would yield an achromatic microscope objective. The flint lens, with its higher index of refraction and greater dispersion, was ground with curvatures just sufficient to compensate the chromatic and spherical aberration introduced by the crown lens.

But fashioning compounded achromatic lenses, consisting of multiple compensating components, proved to be much more difficult for microscopes than it had been for telescopes. The small dimensions of most microscope lens systems made it very hard to construct the components in such a way that the respective aberrations of the two glasses would cancel. There were some successes around 1800 or so in the construction of lower-power doublets (achromatic pairs of crown and flint lenses) and in achromatic microscope oculars; in these cases the dimensions were larger and the desired effect was correspondingly easier to achieve. Higher-powered achromatic doublets and achromatic microscope objectives remained out of reach. Yet hope springs eternal, and most makers of microscopes continued to believe that what could be done in the telescope could be done in the microscope – even if in a modified form.[4]

Among the best known achromatic instruments in Germany were those of Joseph Fraunhofer, a Bavarian optician and maker of some of the most successful achromatic telescopes of the early nineteenth century. Key to Fraunhofer's telescope objectives had been his specially formulated glass (the recipes for which he kept strictly secret), and the empirical methods which he developed for testing the refractive and dispersive properties of those glasses on lines in the solar spectrum. Fraunhofer tried to adapt these methods to the construction of microscopes as well, but they never attained the superior achromatic qualities or the renown of his astronomical instruments.[5]

Achromatic microscopes from the instrument-making Chevalier family in Paris were far more successful in the estimation of contemporary microscopists. After 1823 Chevalier began to sell microscope systems that combined achromatic doublets to produce higher magnifications, although the image quickly degraded when the magnification became too high. Fresnel praised Chevalier's instruments in a report to the Paris Academy, but also said that for many uses non-achromatic instruments were still preferable.[6]

An improvement to the Chevalier approach that was to prove enormously influential was found around 1830 by J.J. Lister, a London wine-merchant and amateur microscopist, who while experimenting with a set of doublets he had ordered from Chevalier, began testing pairs placed in combination. Lister found that there appeared to exist two points, which he named "aplanatic", at which one could place the object for which the doublet remained achromatic and possessed a minimal spherical aberration. Furthermore, he seemed to find that doublets could be effectively combined by causing the aplanatic points from two or more doublets to coincide, resulting in a high-powered, achromatic, spherically corrected microscope objective.[7] This construction, with a series of doublets in the objective, was quite unlike any telescope objective in regular use, and Lister's result apparently surprised optical craftsmen, theoreticians, and microscopists alike. His recommendations at first were largely ignored, but after working for several years to construct aplanatic instruments on his own, in the late 1830s Lister arranged for a few to be produced by different London opticians, who soon began to see advantages in the construction.[8]

Microscope objectives consisting of systems of achromatic doublets (and sometimes triplets) – either removable or fixed – quickly became the norm in the more expensive compound microscopes. The aplanatic points could be found empirically, and the ease of use of the compound microscope could finally be combined with high power and correction of optical defects, both of which had long been available to users of simple microscopes. Compound microscopes rapidly increased in popularity during the 1830s, replacing simple microscopes as the instrument of choice for virtually all types of practitioners.

Lister's work, like almost all experimentation with microscope design, was purely empirical and untainted by mathematical optical theory. This separation between mathematical and practical optics was widespread. Virtually all of those engaged in the construction of optical instrumentation by profession in the early and middle nineteenth century were tradesmen with training in metalwork and the construction of small machines who had picked up their optical skills after gaining employment in a shop engaged in building and selling optical instruments.[9]

In contrast to most of his contemporaries, one microscope maker, the Jena mechanic and optician Carl Zeiss, received some rather atypical training by virtue of the relationship between his *Lehrherr* and the University of Jena. Zeiss had apprenticed to a mechanic in Jena named Friedrich Körner, whose career had been aided by an acquaintanceship with Johann Wolfgang Goethe. It was apparently Goethe's plan that Körner should lecture on practical geometry at the university in Jena, despite his lack of higher educational qualifications. With Goethe's encouragement and support, however, Körner was able to secure a promise from the Grand Duke of Sachsen-Weimar to pay the cost of a university *Promotion*. With the fees taken care of, Körner then petitioned the University Senate to accept "some fragments from my experience written some years ago [*ein seit mehreren Jahren geschriebenen Bruchstück meiner Erfahrungen*]" concerning "refraction by lenses in telescopes" in fulfillment of the requirements for an advanced degree. It was indeed accepted, and Körner was awarded the title of "Doctor" and permission to teach at the university. Though this might appear to have been rather out of the ordinary, Jena already had precedent for awarding "practical men" university credentials and permission to advertise their services through the university's list of lectures. At the same time as Körner, at least two other mechanics without or with limited higher education were listed as providing services and instruction in the making of scientific, surgical, or mathematical instrumentation.[10]

The close connection between Zeiss's master and the university provided the young man with unusual opportunities. Zeiss was permitted to attend lectures at the university while working in Körner's shop, and he apparently did so throughout his apprenticeship. Upon completion of his contract, Zeiss was also able to secure a leaving-certificate [*Abgangszeugnis*] from the university listing the lecture courses he had completed – and these included algebra and analytic geometry, experimental physics, physical and spherical trigonometry, anthropology, mineralogy, and optics (the latter taught by his *Lehrmeister*, Körner).[11] The next several years Zeiss spent as a journeyman mechanic at several different shops in Germany and Austria. There is also some evidence that he attended more lectures and even sat an examination at the Polytechnicum in Vienna. Finally, in 1846, Zeiss returned to Jena to establish a small shop. By this time Körner had long since passed away, and had been succeeded by another mechanic who took over the title of *Universitätsmechanikus*. Zeiss, therefore, became one of three instrument makers practicing his trade in Jena.[12]

Selling eyeglasses and doing small repairs accounted for the bulk of Zeiss's business, but he also managed to sell the odd telescope and construct the occasional small, simple microscope. Zeiss had gotten involved in microscope making at the suggestion of the Jena botanist Matthias Schleiden during the first year or so of his business. At first Zeiss lacked the skills and equipment necessary to cut and polish his own lenses, so he bought them from another optician practicing in Rathenow. In these early years, Zeiss typically sold between 25 and 50 microscopes annually, compared to the 100 or more sold each year by Oberhauser, a leading craftsman in Paris.[13]

Making compound microscopes was at first completely out of the question for Zeiss, but by the mid-1850s he had built up sufficient capital and experience to begin. In 1847, Zeiss took on the first apprentice of his own, one August Löber, who stayed on with Zeiss as a journeyman after completion of his contract, and who in fact remained with the Zeiss company for the remainder of his career. Löber had a knack for the optical work, and within a few years an implicit division of labor had developed, with Zeiss remaining in charge of the business affairs and all mechanical work (such as the construction of the microscope stands) and Löber taking over supervision of all of the optical work (the cutting, grinding and polishing of lenses, as well as the delicate montage, centering and justifying work associated with the assembly of optical components). Löber's work as the skilled craftsman in charge of the optical workshop was absolutely essential in the development and manufacture of each and every one of the products that the Zeiss company produced over the next 40 years. Virtually all associated with the shop in those days agreed that it was his skilled eye that made the difference between top-notch optical performance and blurry junk; it was he who developed the empirical procedures for fashioning and testing the components of Zeiss' optical systems.[14]

Constructing the optical components of the microscope was an art that one had to learn over a long period of time, and that was sometimes extremely difficult to teach. Different employees acquired very specialized functions within the workshop, and when they left often their craft-knowledge would be impossible to replace. For this reason, it was quite common for the proprietor of a workshop to require his employees to sign an oath of secrecy, and Zeiss was no exception. The near irreplaceability of individual employees who had taken up important specialties within the shop is nicely illustrated by a story dating from the 1850s or early 1860s. At the time, Zeiss had only one optician (a man named Rudolf) capable of fashioning the small lenses with tiny radii-of-curvature used as the front-lenses in his higher-power microscope objectives. When Rudolf was called for military service, Zeiss and Löber were unable successfully to train a successor, and were therefore unable to continue to produce the lenses. Faced with a substantial loss of business, Zeiss used his university connections to try to pull some strings and effect a return. He approached the Prorector of the University of Jena with a petition, which was subsequently ratified by the university Senate, requesting the Grand Duke to order Rudolf's release from service. The petition was successful, and Rudolf was marched back to Jena to take up his old position at the polishing-bank in Zeiss's workshop. According to those who claimed to remember the incident, Zeiss was moved to tears by his man's return.[15]

By the middle of the nineteenth century the boundaries that separated the mathematical analysis of optical systems from the practice of actually building those systems were beginning to be crossed with increasing frequency and with increasing success in other areas of optical craftsmanship such as telescope and camera making. Collaborations between Ludwig von Seidel and Carl August

Steinheil in the design and manufacture of photographic objectives,[16] the work of Vienna mathematician Joseph Petzval on the portait objective,[17] and collaborations between the Voigtländer family and a number of Viennese mathematicians (including Petzval) on both telescopes and cameras served as powerful examples of what could be done in the manufacture of certain instruments when practical skill and mathematical analysis were combined.[18] During the 1840s, '50s and '60s, collaborations between academically trained mathematicians and craft-trained instrument makers became more and more common, but the microscope seemed to remain outside the reach of formal optical theory.[19]

At mid-century a large proportion of those with mathematical and physical training who pursued problems in geometrical optics were astronomers by profession and unfamiliar with the use of the microscope. A position in an observatory had long figured in the career paths of men with mathematical and physical training (unlike, for instance, positions at botanical gardens), and a large proportion of those who held university chairs in physics had or continued to have a professional relationship with an observatory.[20] The neglect of the microscope was noticed by those who *were* interested in its use. The botanist Matthias Schleiden, for instance, stated in a piece devoted to the role of the microscope in biological investigation: "Unfortunately in the writings of the physicists one finds not a single bit of information, because to this date not a single one has concerned himself with the theory of microscopic investigation."[21] He was not far off.

This lack of serious theoretical treatment of the problems of microscopy would soon change, however, with the emergence of several new schools in animal and plant physiology interested in approaches involving the application of mechanics to the understanding of vital processes. They approached biological problems using the tools of chemistry and physics, and they insisted on the mathematical description of the processes and structures they studied.[22] And, as we shall see, they also insisted on a mathematical description of the microscope, their chosen experimental tool for the reform of biology.

SEEING IS NOT BELIEVING: MICROSCOPISTS CONFRONT THE MICROSCOPE

As of 1830 or so the microscope was still regarded in continental scientific circles largely as a curiosity that had no real place in serious scientific work. Many shared the opinion of Xavier Bichat and the followers of his school of "general anatomy", who believed that microscopical observation was irrelevant to the study of living organisms. And even setting programmatic objections aside, there already existed many microscopical observational accounts containing improbable findings of "globules" and other structures in living tissues that later were discredited as artifacts of poorly constructed or poorly deployed instrumentation – most likely the result of excessive magnification together with uncorrected aberration. To many, the microscope seemed an untrustworthy instrument indeed.[23]

This distrust of the microscope began to change among German and central European biologists during the 1830s as they began to use the new and improved achromatic microscopes that were rapidly becoming available. Jan Evangelista Purkynu, physiologist at the University of Breslau, for instance, undertook a series of microscopic studies that greatly influenced his contemporaries in which he examined the intracellular fluid, calling it "protoplasm" and suggesting that it might be the most basic material of life. Johannes Müller in Berlin, who had previously found the microscope to be of limited use at best – he had at one time argued against including its use in the regular training of physicians – began microscopical investigations himself and also began to teach his students the techniques of microscopy. Others soon followed the example.[24]

Many of Müller's students became advocates of what collectively came to be known as the cell theory – the doctrine that the tissues of living organisms are built from discrete living units – and it was in their work that microscopical techniques were to assume a new, central role. For the cell theorists, the microscope was to become the instrument that defined their work, for it made visible the structures that they regarded as crucial for the proper understanding of the organism. Reservations about the reliability of the instrument, however, remained a barrier to the acceptance of their program and their findings.[25]

In order to counter these objections, by the early 1840s several biologists were calling for texts on the theory and proper practice of microscopy. Their aim was to describe what they regarded as proper microscopic practice, in order that the work that seemed to discredit the instrument could be dismissed as the results of poor method. Furthermore, since some of their opponents were prepared to argue that the microscope could not – even in principle – yield reliable, replicable results, they believed that a thorough treatment of the physics of the microscope would help legitimate the practices they advocated.

Matthias Schleiden, a former student of Müller's and one of the authors of the early cell theory, issued one of the earliest calls for a firmer foundation upon which to build rules for the proper practice of microscopy in his *Grundzüge der wissenschaftlichen Botanik* of 1842. The book, Schleiden's manifesto for the future of botany, contained a lengthy "methodological introduction" outlining the proper practice of the study of vegetable life and went through four editions over the next twenty years; it was one of the main botanical textbooks in Germany up to the 1860s.[26] At the center of Schleiden's methodology was the microscope:

What would one say to a cabinetmaker who delivers clumsily made and unusable goods, yet excuses himself by saying that he has no lever or saw? "My friend, become a shoe-cleaner or whatever else you want, for God's sake, but you cannot and never will be a cabinet-maker". *Fiat applicatio.* Of course, there are only a very few who imagine themselves to be able to accomplish something without an indispensable tool of the trade. One finds that most botanists possess a more or less worthy [microscope].[27]

Schleiden sought a "theory" of the microscope to give a quantitative account of the properties of the image that the instrument produces in order to show that there could be no reason why properly executed microscopical observation need be any less reliable than observations performed with the naked eye:

Our eye is itself an optical apparatus, as we have already seen; the microscope uses almost exactly the same means, and we must first and foremost understand that the microscope can never give us different quality than does the eye. . . . On the other hand, we must understand that the functioning of the eye, so long as it is healthy, follows the same exceptionless mathematical laws as does the microscope, and that therefore in all observations using the naked eye or using a microscope only one's judgement can err – one's healthy senses and the optical instrument are always correct.[28]

Though there may be numerous sources of aberration and false appearances in the microscope, they could be combatted with proper practice justified on the basis of optical theory, and erroneous results could be dismissed as the products of shoddy practice. For Schleiden, the role of optical theory was to tell the microscopist when he could draw sound conclusions on the basis of what he saw.

Another early advocate of one of the many cell theories who also placed the microscope at the center of his plans for reform of the life sciences was Rudolf Virchow (though his view of the cell and his plan for its investigation were quite different from Schleiden's). Virchow wrote: "It is necessary for our conceptions to advance at least as far as our ability to see has been extended by the microscope: all of medicine must step 300 times closer to its natural predecessor" [i.e. the observation of natural objects]. But if microscopic observation was ever to overcome the widespread skepticism against it, he argued, treatises were needed that could advocate proper microscopical practice, and which would prove on the basis of optical theory that proper practice would lead to valid observations:

Though from the beginning received with a great and general admiration, [the discoveries of the cell theory] have gradually lost their credit with a large number of physicians, all the more because in their partly justified mistrust, these physicians have found comfortable reasons for keeping the microscopists at a certain respectful distance and to assign to them a humble place in the audience-room or in the clinical Cortège . . . The Wochenschrift für die gesamte Heilkunde writes the word "microscope", whenever it is forced to mention it in one of its epigrammatic criticisms, with an exclamation-point, and one occasionally hears a young practitioner with a dismissive gesture exclaim, "So it's microscopic, is it?"! . . . Because they do not know anything about the microscope themselves, they cannot know that the number of optical illusions are small and that the majority of these are actually logical illusions – false interpretations of objects correctly seen.[29]

For Virchow, as for Schleiden, it was absolutely necessary to convince the scientific public that the microscope was a suitable instrument for natural philosophy. Unless that argument could be won, there would be no future for cellular pathology or the medical reform that its supporters advocated.

The legitimation of microscopical practice in the study of living organisms was often a stated goal in texts about the microscope that appeared in Germany during the 1850s and 1860s – a very large number of them written by students of Schleiden or of Virchow and carrying discussions of micro-scopical optics to varying levels of detail.[30] Microscopists defended their enterprise with new forms of microtechnique, staining, and other manipula-tions, but optical arguments remained prominent.[31] These texts on the use of the microscope contained virtually the only attempts to work out a geometrical optics for the instrument between the 1830s and the 1870s.

The first treatises that appeared in response to the calls of Schleiden and Virchow emphasized the importance of proper observational technique – techniques that respected the optical qualities of microscopical vision. Technique was paramount, but theoretical justification was needed as well. Theory had to show how the balls, blobs, rings, halos and general fuzziness could exist and still not threaten the basic premise of microscopy: that, properly interpreted, the microscopist could believe what he saw. Most of these texts began with a consideration of how images are formed and perceived by the unaided eye, emphasizing how various conditions such as brightness, dimness, color, size and distance affect appearances. From there, they tried to show the similarities and differences between normal and microscopical sight.

One pressing problem for the practicing microscopist was to explain why differing illumination techniques seemed to produce wildly different results for different objects. This particular question figured importantly in later debates about the function and theory of the microscope. Some objects seemed to be viewed to best advantage under direct illumination (with the light source behind the object, on or near the optical axis), while others were best seen when in oblique or frontal illumination. And worse: observers did not always agree about the best technique for any particular object. There seemed to be no rules that could be applied routinely.

For example, in order to account for the problem of illumination, Pieter Harting undertook an empirical investigation, instructing a number of skilled observers to perform various exploratory trials to test their perceptions in several controlled circumstances. Harting found that the eye could see very small positive images (light on dark background) much more easily than it could negative images (dark on light background). He also found that the shape of the object seemed to make much more difference in perceptibility when the image is negative than when it is positive; dark, thread-like objects are much more easily seen than dark, round ones when placed against a bright background but bright thread-like and bright round images placed against a dark background seemed just as difficult to see. Intensity of illumination, on the other hand, seemed to make much more of a difference with positive images than with negative.

The type of illumination, whether reflected from in front of the object [auffallende] or transmitted through it [durchfallende] also seemed to play an important role in determining the perceptibility of the object. For reflected light, the shape of the object and its coloring (more precisely the relation between the color of the body and the background) were the most important factors in determining the limits within which an object can be seen. But for transmitted light, a number of other factors came into play; generally speaking, the object would be easier to see, said Harting, when more rays were diverted from the path that they would have taken to the retina, had the object not been in their way.[32]

Illumination was just one technical problem confronted by microsopists. Other problems stemming not from user's techniques but from instrument design also merited detailed attention. Their insistence that no theory could instruct the optician did not mean that they believed opticians created no bad instruments. A consideration of different microscopes from different makers and of the different optical properties and mechanical conveniences that they possessed was a prominent feature in most of the microscopists' treatises. Again, the purpose was to show that the instruments *actually available* were understandable, predictable, and worthy of consideration as scientific instrumentation.

The relationship between aperture size and image sharpness was a central issue in accounts of the relative merits of particular microscope systems. Increasing angular aperture was generally considered to be an important parameter in determining image clarity, yet significant differences of opinion persisted about the cause of that connection. Along with debates about illumination, controversy about the nature of the relationship between aperture and sharpness, closely related to debates about the effects of different types of illumination, came to be at the very center of disputes about theories of microscopical image formation and about the function and value of new microscope technologies of the 1870s and 1880s.

In trying to explain how it was that wider apertures presented finer details, analogy to telescopes proved useful. All of these authors followed to some degree language introduced by the British physician C.R. Goring in the 1830s, applying to the microscope the distinction between the correction of aberrations and the ability to make distant objects visible that was already long familiar from William Herschel's discussions of the telescope. In this vocabulary, "defining power" referred to the absence or presence of geometric or chromatic aberration while "penetrating power" referred to the qualities of sharpness that seemed to come with wider apertures. According to Goring, defining power (or the quality of the correction of aberrations) was to be tested by looking for the sharpness of edges and was primarily dependent on the correction of spherical and chromatic aberrations, while penetrating power (the effect of aperture) was to be tested by attempting to view very small objects, very small details on objects, or very finely spaced lines.

Even though most microscopists were convinced of the relationship between aperture and image quality and even though virtually all of the German authors on the subject adopted Goring's terminology, the cause of the relationship was a source of confusion.[33] Although the relationship between sharpness and aperture in the microscope may have been found originally on the basis of an analogy with the telescope, warned Harting, the analogy could not be carried over into causes, as Goring's vocabulary might suggest. Because the two instruments were used to form images of very different kinds of objects, the conditions that affected how these objects were perceived were also quite different, and the rules of practice in astronomy and in biology could not be the same. And the different conditions of visibility indicated that different causes were at work:

One must not lose sight of the fact that these two instruments are only completely comparable when they are both used to form positive [light on dark] images. That is the case in the telescope when it is turned on luminous objects in the heavens and in the microscope when one observes an object using incident [auffallendem] light. I need require no proof that the visibility of such objects increases with the brightness of the image, and therefore with the size of the aperture angle of the objective.[34]

But in the case of the microscope penetrating power should also refer to the ability of the observer to make out details on the surfaces of objects that are at least partially transparent, details like the markings on insect wings and tiny diatoms. Since rays that encounter a boundary between different optical media at a greater angle are reflected or refracted at a similarly greater angle (as in the discussion of the resolution of lines in a grating), this makes the difference between the lighter and darker parts of the regions under oblique illumination or when using wider aperture even more pronounced.

Leopold Dippel disagreed with Harting and argued that brightness was very much the point. Dippel argued that all microscopic images were in some way combinations of transmitted and reflected light. When the image is formed from light passing through a semi-transparent object, as is the case in the microscope most of the time, one must consider the image partly as formed by the deviation of rays by differences in the optical properties in the media within the object, and partly as formed by rays that pass through undeviated, as if the object were self-luminous. Therefore, said Dippel, Harting was wrong to deny the role of the increased brightness in accounting for the improved penetrating or resolving power that accompanies wider aperture.[35] The relationship between aperture and image quality, while seemingly undeniable, was also contentious and not entirely explicable. It was, maintained all of these authors, nonetheless one of the first properties of the instrument with which the microscopist must come to grips – especially before purchasing a set of objectives.

The penultimate treatise on microscopical optics produced by this generation of microscopists appeared in 1867. Das Mikroskop, written by cell theorists Carl Nageli and Simon Schwendener, contained one of the most thorough and mathematical treatments of the optics of the microscope to date, and considered critically the characteristics of microscope performance, including the relationship between aperture and sharpness. Nägeli and Schwendener argued that widening the aperture of a telescope has precisely the same effect as widening the pupil of the eye – more light, i.e. a cone of greater solid angle but equal intensity, is admitted. When the light coming from the object is rather weak, as in the case of the telescope, this can have a beneficial effect by making more light available to form a brighter image and allowing smaller, dimmer objects to be seen. But when the light coming from the object is more intense, they said, more aberration due to the wider aperture may overwhelm any benefit from increased gathering of light, which may itself no longer be necessary. In the case of the microscope, they pointed out, the intensity of the illumination is under the control of the observer and one can usually bring to bear as much light as one chooses, even using direct sunlight when necessary to achieve great brightness and intensity without widening the aperture of the objective. Furthermore, they argued, "as everyone knows" the stops generally used in the illumination apparatuses of microscopes are so small as to limit the aperture angle of rays converging on points in the object from the source of light to a mere 20°–30°. Although the light-cone would be broadened somewhat by refraction in the object, they argued that this would rarely increase the width of the light-cone by more than a few degrees. Therefore, most of the opening in the higher-aperture objectives (60° and higher) actually went unused. Widening the aperture of the microscope objective did not increase the amount of light coming into the microscope when the illuminating pencil was limited in this way (which they believed was typical) and should have no effect on the properties of the image at all! Wider aperture, they concluded, must be a side effect – a symptom, not a cause of greater penetrating power.[36]

The origin of the observed relationship between aperture and sharpness therefore had to be found in some other property of the objective. Perhaps, they suggested, optical craftsmen were somehow able to create better aplanatism when the apertures were wide, and this effect had mistakenly been attributed to the aperture itself. As we shall see, Nägeli and Schwendener's discussion of aperture and its lack of effect on the images formed by the microscope, together with the subsequent empirical determination that large parts of the objective did, in fact, remain dark when moderate illumination pencils are used, turned out to be enormously important during the 1870s in shaping the development both of the theory of the microscope and the evolution of optical design.

ERNST ABBE AND THE PHYSICS OF THE MICROSCOPE, 1865–1875

The physicist Ernst Abbe would do more than any other individual to make academic optical theory relevant to the construction and design of the microscope. In collaboration with Carl Zeiss and the craftsmen in his workshop, during the 1870s and 1880s Abbe reinvented the process of the design and construction of microscope systems, introducing to the fabrication of these kinds of optical instruments precise control over raw materials and system parameters, new methods of designing systems with the aid of calculation and optical theory, and new optical materials delivered with precisely determined properties. Abbe was also the author of a theory of image formation in the microscope which came to be adopted by many microscopists as the basis for normative microscopical practice. Ironically, it showed that the early fears about the inherent unreliability and instability of microscopic evidence were well founded. But by telling observers what they could not trust, Abbe's theory also delineated what they could rely on.

Ernst Abbe was trained as a physicist, with a strong emphasis on mathematics, and had essentially the same tools at his disposal, both theoretical and experimental, as any of his colleagues. Having studied for two years at Jena and three more at Göttingen, Abbe's main area of specialization had been "critical or knowledge-theoretical" [kritisch-erkenntnistheoretische] (as opposed to "experimental") physics, which he had learned primarily under Wilhelm Weber and Berthold Riemann.[37]

Even though experimental physics was not his main interest, experimentation was still a major part of Abbe's training. At Göttingen, the training of a physicist included a heavy dose of the theory and practice of precise measurement. Much of this training came in the physical-mathematical seminar led by Weber, whose main topic, Abbe reported, was the theory of measuring instruments (with particular emphasis on the precise measurement of weak electric currents and magnetic fields). It was quite common for Weber to use the seminar to train observers in the techniques he used in his own experiments on terrestrial magnetism, employing students in observations as they became proficient.[38]

Students at Göttingen typically recieved hands-on training in the design and construction of measuring instruments, often in the workshop of Moritz Meyerstein, Göttingen's *Universitätsmechaniker*. The majority of students who studied the natural sciences were bound for careers teaching secondary school, and the ability to fashion the apparatus for one's own demonstrations was considered to be essential. One of Abbe's closest friends and fellow physics student, for instance, took a course in Meyerstein's shop in turning, filing and glass-blowing, part of his preparation for the *Oberlehrerberuf*. Even Abbe, who had decided early on that he had "damned little interest" [*verdammt wenig Lust*] in school teaching, nevertheless spent time in Meyerstein's shop, helping him with the design of a new spectrometer and an electrogalvanometer after completing his regular studies. Most of Abbe's work for Meyerstein involved the analysis and improvement of the precision of his instruments. A topic in the mathematical analysis of precision and experimental error was, a few years later, also the subject of Abbe's *Habilitationsschrift*.[39]

Optics was not a major part of Abbe's training, nor was it a subject in which he showed any special interest. Having taken a course of lectures and laboratory exercises in "analytical optics" during his third semester at Jena, Abbe also took a course in optics from J.B. Listing in Göttingen, though Listing's emphasis in the course, as in his own research, was strongly physiological. Not much interested in the optics of the eye, Abbe found Listing's course rather less compelling than Weber's seminar on the theory of measuring instruments. His first intensive exposure to the details of optical instrumentation came a few years later, while working as an assistant at the Göttingen observatory. But his duties there had to do with astronomical observations, not optical instrument design.[40] Abbe was no specialist in optics when he returned to Jena in 1863 to complete his *Habilitation* and begin a career as a university instructor.

PHYSICISTS DISCOVER THE MICROSCOPE

Several years after Abbe began as a *Privatdozent* in Jena, Listing briefly turned his attention to the microscope, and became one of the first physicists to

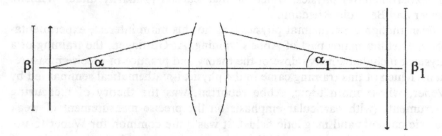

Figure 1. Diagram for Helmholtz's calculation of the relationship between brightness and magnification. α and α_1 are the divergence angles in the object and image spaces. β and β_1 are the linear dimensions of the image and object surfaces perpendicular to the axis. When the optical system is a microscope, α is one half the angular aperture. Adapted from Helmholtz (1874), 561.

publish work on the particular optics of microscope systems. Listing thought that he could boost the magnification that a microscope could deliver while simultaneously increasing its defining and penetrating power by using a technique adapted from terrestrial telescopes. Telescopes made for terrestrial use generally involved the formation of two successive real images within the instrument (one between objective and ocular, the second within the ocular) so that the image as viewed by the observer would stand upright. Listing believed that a similar arrangement would have advantages for the microscope, perhaps yielding useful magnifications as high as 25,000 or 50,000 times, because each successive real image could be magnified using a relatively low-powered system, reducing the amount of aberration introduced. The strongest contemporary microscopes on the continent advertised maximum magnifications on the order of a few thousand times.[41]

Listing's claim, published in a physical journal, apparently led Hermann von Helmholtz, another physicist whose best known optical researches had been in the area of physiological optics, to consider whether there might be a physical limit to the efficiency of the microscope. Listing's suggestion seemed not at all to agree with the experience of practitioners that improvements in performance could be won only after much effort. It also seemed at odds with the practical experience of microscopists that penetrating power increased primarily with larger angular apertures, not with changes in other parameters of the microscope.

Helmholtz argued that decreasing brightness with increasing magnification probably limited the degree of magnification that could usefully be employed in a microscope. He derived an equation relating the divergence angle of the object and image pencils on either side of the optical system (See Figure 1), and then used this relation to consider the case of an image emerging from the instrument not large enough to fill the pupil of the eye of an observer. In this

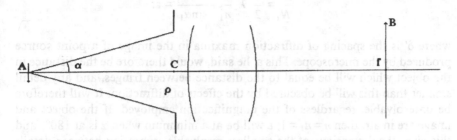

Figure 2. Diagram for Helmholtz's calculation of the effects of diffraction in the microscope. Again α is half the angular aperture of the system. C is the center, located on the optical axis, of the stop with radius ρ which limits the passage of rays into the system. Adapted from Helmholtz (1874), 581.

case, the brightness H of the image is always less than the brightness H_0 of an image falling on a relaxed eye filled with light.

The magnification N_0 at which the light pencil emerging from the instrument exactly fills the pupil will be reached when $H = H_0$ and is given by

$$N_0 = \frac{s}{p}\sin\alpha_0.$$

For fixed aperture angle a_0, $H:H_0 = N_0{}^2:N^2$ when $N > N_0$ and N is the magnification of the system. N_0 will increase proportionally to the sine of the aperture angle, and will reach its maximum when that sine is equal to 1. So, argued Helmholtz, if s is taken to be the conventional 250 mm and p for bright illumination is 1.5 mm, then the highest achievable N_0 is approximately 167. At double this magnification, the brightness of the image will fall to one fourth of its value at N_0; at four times this magnification it falls to around a sixteenth. Drastic increases in magnification will create an image that is too dim to see, and magnifications like those hoped for by Listing would be unusable.[42]

But even if this difficulty could be overcome, argued Helmholtz, diffraction effects limited the size of the objects visible in a microscope. Because the diameter of the front lens of a typical high-powered microscope is so very small, the opening of the objective affects light passing through it much as a pin-hole does, and thus causes diffraction fringes to appear around the image of a light source. Consider light coming from point A on the optical axis, passing through an opening (i.e. of the microscope objective) with its center at C and with radius ρ (see Figure 2). Using formulae for the spacing of diffraction maxima caused by light from a point source A passing through a small circular opening (c) together with a result from the first part of his analysis, Helmholtz got

$$\frac{\delta'}{N_1} = \frac{1}{2} \lambda \frac{n}{n_1} \cdot \frac{1}{\sin\alpha_1} = \varepsilon.$$

where δ' is the spacing of diffraction maxima in the image of a point source produced by the microscope. This ε, he said, would therefore be the distance in the object which will be equal to the distance between fringes, and any detail smaller than this will be obscured by the effects of diffraction. It will therefore be unresolvable, regardless of the magnification employed. If the object and image are in air, then $n = n_1 = 1$; ε will be at a minimum when α is at 180°, and this gives the dimensions of the smallest resolvable separation between details achievable using a non-immersion microscope:

$$\varepsilon = \frac{1}{2} \lambda$$

Helmholtz concluded that craftsmen could make instruments which approached, but could never exceed, this fundamental limit.[43]

ABBE AND ZEISS

Just as Helmholtz was finishing his manuscript, he came across an article published several months earlier by the young physicist Ernst Abbe. Abbe had reached conclusions very similar to Helmholtz's. Because his article was to be part of a special celebratory issue of the *Annalen der Physik*, Helmholtz did not feel at liberty to pull it, and instead attached a brief afterword acknowledging Abbe's work and his priority.[44]

Abbe had come to his researches on the theory of the microscope after several years of collaboration with Carl Zeiss. Abbe apparently approached Zeiss concerning the construction of apparatus for demonstrations that were to accompany his lectures on the theory and use of measuring instruments. Hoping to reproduce at Jena the kind of lectures and exercises in which he had taken part at Göttingen, he wanted to have an apparatus built for the precise measurement of weak electrical currents and magnetic fields based on Weber's methods. Zeiss was unfamiliar with this kind of instrument, and Abbe had to take an active role in overseeing its design and construction in Zeiss' shop.[45] The two men got along well. A few years later when financial difficulties forced Abbe to consider finding additional income to supplement his lecturing fees, Abbe began working for Zeiss, much as he had for Meyerstein in Göttingen, on other projects.

During these early years with Zeiss, Abbe did what a trained physicist with his background did best – he designed measuring instruments. It had always been a source of frustration to Zeiss, who delegated virtually all responsibility for the optical work to his colleague August Löber, that the optical components of his systems could not be designed and fabricated according to specifications in the same way as the mechanical components like the stage or

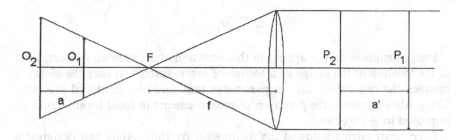

Figure 3. Schematic diagram of Abbe's focometer. His method used images of two different objects, scales ruled on glass, at P_1 and P_2 and a determination of the magnification produced by the system of those two objects to determine the focal length f. Adapted from Czapski (1892), 189.

stand could. The instruments that Abbe designed held out the promise of allowing Zeiss more precise and more direct control over the manufacture of those optical components.

The first instrument Abbe built for Zeiss' shop was a focometer, used for measuring the equivalent focal lengths of individual lenses, lens combinations, and completed systems. Most methods at the time involved using the system to be measured to project an image of an object of known size against a screen and deduce the focal length from the distance of the image and the magnification it had undergone. Abbe's analysis showed that there were tremendous sources of error in this method arising from uncertainty in determining the location of the image (it was affected by the sharpness of vision of the person doing the measuring and the aperture of the system being measured); small changes in the adjustment of the apparatus could sometimes have large effects on the size and location of the image. Abbe also found that distortion near the edges of the image could affect the determination of the magnification of the image. Together, these factors led to errors of the order of 5–10%.[46]

Abbe devised a new method which used precisely ruled scales as objects so that distortion and edge effects would not obscure the measurement. He also used object, rather than image distances by requiring determination of the magnification brought about of two different objects O_1 and O_2 at different, known distances from the system. (See Figure 3.) If f is the focal length of the system, N_1 the magnification brought about between a pair of conjugate points on the axis, N_2 the magnification between two other conjugate points, and a the distance between the two object planes, then the focal length of the system is

$$f = \frac{a}{\dfrac{1}{N_1} - \dfrac{1}{N_2}}$$

Image distances do not appear in this equation at all, removing uncertainty in the location of the image as a source of error. Instead, it uses the distance between the two objects (a) – a parameter that could be measured precisely. Using Abbe's system, the precision of measurements of focal length could be improved to ±1% or so.[47]

Other instruments followed the focometer. By 1869, Abbe had designed a new type of refractometer for measuring the indices of refraction of solid and liquid bodies. Here, as with the focometer, his goal was to increase the precision of the measurement, and to make the apparatus usable by non-specialists:

The method first taught by Fraunhofer, according to which the minimal deviation of the rays of each specific color are observed in a prism of known angle, fulfills without doubt all of the requirements that one would want to make of it for various goals, so far as the accuracy and precision of the results are concerned ... On the other hand, it should not be overlooked that measurements of this kind remain a complicated and subtle business, and only an experienced observer is truly equal to the task. ... In fact, Fraunhofer's method has not spread widely beyond the physical laboratories. The practicing optician, now as before, typically employs a number of much simpler, but also much inferior, procedures ...[48]

Abbe's method used total internal reflection within a prism made from the glass whose index was to be measured to cause an incident beam of light to retrace the path to its source. When the image of the light source and the source itself coincided, the observer could read the parameters necessary to determine the index of refraction from a scale. These instruments were much easier to build than the usual spectrometer, and very much easier to use. The same method could also be used to measure the index of refraction of fluids. (See Figure 4.)

An apertometer for measuring the angular aperture of microscope objectives followed – designed and in use by sometime in 1870.[49] A comparator, for precise measurement of distances, with variations of the instrument for measuring the thickness and radii of curvature of lenses also dates from approximately this period (see Figures 5,6,7).[50]

Abbe worked closely with both Zeiss and Löber to integrate measurement and testing procedures into the fabrication process. Other, more qualitative tests were also employed. One of these, developed by Löber, allowed precise control over the sphericity of the lens surface. Löber would place a lens to be tested directly on top of a pattern-lens with the same but opposite radius of curvature (e.g., if the lens to be tested was convex, the pattern-lens would be concave) and with the help of a small microscope would look for Newton's rings. Irregularities in the rings would indicate deviations from sphericity, and these would be removed by further working of the lens.[51]

With the tools to measure the parameters of Zeiss' objective systems, and with a possibility to ensure (to an extent) that the fabricated systems would conform to specification, Abbe began to investigate the geometry of Zeiss' systems and components and to begin to calculate the path of rays through them with a view to improving their performance. The first of these calculations date from around February 1869, after he and Zeiss had been working together for almost three years.

Abbe's early attempts at calculating microscope objectives were miserable failures,[52] but his designs improved as he began to realize the importance of the correction of aberrations off of the optical axis. These aberrations are of relatively little importance in instruments like the astronomical telescope, where the objects can be treated as infinitely distant point sources and where the angular aperture of the system is negligible, but become sizable under the conditions that prevail in the microscope. Especially crucial to the change in Abbe's fortunes seems to have been expanding on a requirement for the correction of spherical aberration – that rays which enter the objective parallel to the optical axis be united in a single focal point. By further requiring that such rays also yield images of precisely the same magnification (i.e., requiring rays entering the system at different distances from the optical axis to be imaged using precisely the same focal length), Abbe could ensure that extended objects, not just point sources, were imaged without aberration. This requirement came to be known (in a somewhat more general form) as the "Abbe Sine Condition".

If the sine condition could be met (and ordinary spherical aberration corrected) for rays of all colors, the resulting image would be completely free from aberration. With these criteria, it now became possible for Abbe to calculate ahead of time – without further empirical testing – whether a particular change would improve or worsen the convergence of rays by the system. The new techniques of measurement and control now available in Zeiss' shop gave some assurance that the designs Abbe created on paper could be realized in practice.

Abbe began employing a version of the sine condition sometime in late summer or fall of 1870 and by the end of 1871 some improvements based on his calculations were, with Löber's help, beginning to find their way into Zeiss' products. Zeiss now sensed opportunity, and in October 1871 he entered into separate, formal agreements with Abbe and Löber according to which they would collaborate in the design of new and improved microscope systems and both would receive a share of the revenues that these new systems generated. During the next year and a half, Abbe and Löber worked to redesign and test Zeiss's entire existing line of microscope objectives, and to develop a new set of higher performance objectives to complement it.[53]

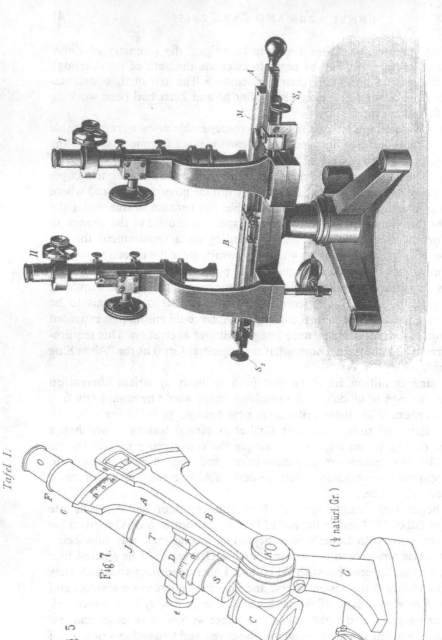

Tafel 1.

Figure 5

Figure 4

Fig. 7

Fig 5

(½ natürl Gr.)

Figures 4, 5, 6, 7: Abbe's instruments. Figure 4: Abbe refractometer for measuring the refractive index of fluids. A small amount of the liquid would be placed between two prisms and the angle determined at which total internal reflection occurred in the liquid. Figure 5: Abbe's comparator. Figure 6: Abbe's thickness gauge. Figure 7: Abbe's spherometer. The drawing of the refractometer comes from F. Abbe.

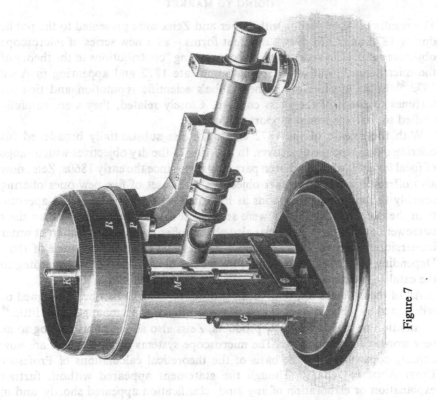

Figure 7

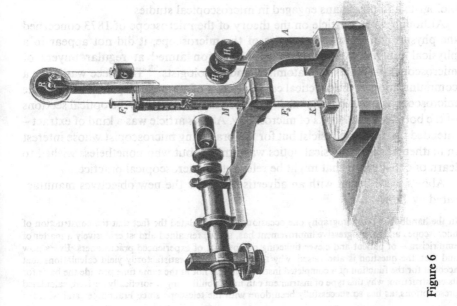

Figure 6

GOING TO MARKET

The results of Abbe's labor with Löber and Zeiss were presented to the public during 1872 and 1873 in two different forms – as a new series of microscope objectives and as an article by Abbe concerning "contributions to the theory of the microscope", sent to the publisher in late 1872 and appearing in April 1873.[54] Both, in time, established Abbe's scientific reputation and tied his fortunes to those of the Zeiss concern. Closely related, they were carefully crafted so that one would support the other.

With the release of his 1872 catalog, Zeiss substantially broadened his offering of microscope objectives. In addition to the dry objectives with a range of focal lengths, which had been part of his line since the early 1860s, Zeiss now also offered a series of weaker objectives and a set of five new ones offering roughly the same magnifications as his usual series, but with wider aperture than the older models. These were sold for a slightly higher price than their narrower counterparts. Zeiss was also able to offer for sale three different water immersion objectives, each advertised as possessing an aperture angle of 180°. Depending on the choice of ocular, the strongest of these would, according to the catalog, yield magnifications between 860 and 2400 times.[55]

All of these objectives, both older and newer models, had been designed or redesigned by Abbe, though the older models had, in fact, changed very little.[56] In addition to the array of new products, Zeiss also added to his catalog some new words of introduction: "The microscope systems produced here are now entirely constructed on the basis of the theoretical calculations of Professor Ernst Abbe in Jena."[57] Though the statement appeared without further explanation or elaboration of any kind, clarification appeared shortly, and in a forum likely to attract the attention of Zeiss' best customers – academic biologists and physicians engaged in microscopical studies.

Although Abbe's article on the theory of the microscope of 1873 concerned the physics and geometrical optics of the microscope, it did not appear in a physical journal but rather in a publication aimed at regular buyers of microscopes – botanists, anatomists, and histologists.[58] The piece was to be a communication of the practical consequences of a more complete theory of the microscope which Abbe planned to publish later. Much like the optical sections of the botanists' textbooks of microscopy, Abbe's article was a kind of extract – intended not for the physicist but for the practicing microscopist whose interest in mathematical or physical optics was limited, but who nonetheless wished to learn of conclusions that might be relevant to microscopical practice.

Abbe's piece began with an advertisement for the new objectives manufactured by Zeiss:

In the handbooks of micrography one occasionally finds related the fact that the construction of microscopes and its progressive improvement has, so far, remained almost exclusively a matter of empiricism – of patient and clever tinkering on the part of experienced practitioners. Every now and then the question is also raised: why theory, which can satisfactorily yield calculations that account for the function of a completed instrument, can not at the same time provide the basis for its construction; why this type of instrument can not be built using theoretically derived, calculated prescriptions, as has so successfully been done with the telescope since Fraunhofer and with the camera more recently... With this in mind, and in collaboration with Mr. C. Zeiss in Jena, I have undertaken an earnest attempt to give the construction of the microscope and its further improvement just as certain a theoretical basis as the manufacture of telescopes was given by Fraunhofer.... For some time now microscope systems have been assembled completely according to theoretical prescription – systems whose performance is comparable to previous ones, from the weakest to the strongest.[59]

In advocating his new "scientific microscope-making", Abbe found it prudent to go slowly. At first he claimed only that calculation and measurement had been used to make objectives that were at least as good as those produced using the traditional methods; later he would build a case that they were better:

The optical constants of test prisms from every piece of glass to be worked were measured so that variations in the materials could be compensated for in the design. The individual components were fashioned and assembled as exactly as possible to the prescribed dimensions, and only in the strongest objectives was a parameter (a lens-distance) left variable so that the small, unavoidable deviations in the work could be compensated. This work shows that a sufficiently fundamental theory in combination with rational techniques, making use of every tool that physics can offer to practical optics, can, even in the construction of microscopes, successfully replace simple empiricism.[60]

Zeiss' objectives had already come to enjoy an excellent reputation in Germany,[61] and given the widespread sentiment among microscope buyers that the design and fabrication of delicate, high-powered objectives was best left in the hands of the artisan – that the use of too much optical theory was likely to do more harm than good to the optics of the objective – the statement that Zeiss was now collaborating with a young physicist by itself probably seemed unlikely to bring in many new customers. In order to turn the announcement from a disclaimer to an advertisement, Abbe had to give positive reasons in favor of bringing the tools of physics to the construction of microscope objectives. He needed to make a case that his brand of scientific knowledge (as opposed to the purely craft knowledge of the optician) would help to create a *superior* microscope objective. His strategy was to redefine what "superior" meant.

A properly scientific evaluation of the performance of a microscope, suggested Abbe, was based on a refined understanding of the process of image formation and, with it, a better understanding of the causes of image imperfections. There are aberration-effects that are purely dioptrical, like spherical and chromatic aberration, and aberration-effects that are not dioptrical, such as diffraction. His finding that non-dioptrical processes play an especially important role in the formation of images of fine details later came to be regarded as the hallmark of the Abbe theory.

The first part of Abbe's paper concerned the "dioptrical conditions for the performance of the microscope", or rather the results of a geometric analysis of the path of light-rays through the instrument. He classified geometric and chromatic aberrations as either errors of focus, or errors of magnification:

To the first group [errors of focus] belong the spherical and chromatic aberrations that one usually considers; the second group [errors of magnification] contains a series of peculiar deviations from the normal path of the rays that together arise from the fact that the homofocal pencils which fill the different parts of the free aperture [i.e., that are incident at different distances from the optical axis] deliver images of unequal magnification according to the degree of different inclination of these

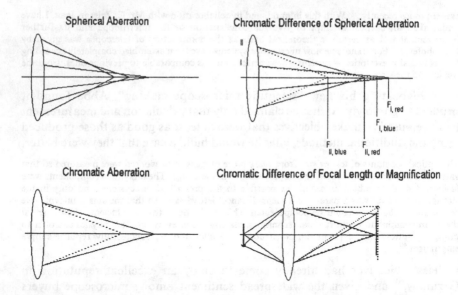

Figure 8. Schematic representation of the main types of geometric and chromatic aberration discussed by Abbe. (1) **Spherical aberration** results from the fact that refraction from a spherical surface does not bring rays to a perfect focus. A convex lens, like the one shown here, is called spherically undercorrected, as rays from the periphery are brought to a shorter focus than rays closer to the center. Concave lenses are spherically overcorrected. (2) **Chromatic aberration** occurs due to the different refrangibility of light rays of different colors, blue being refracted more strongly than red. In the diagram, the dashed lines represent blue rays, which are brought to a shorter focus in an uncorrected system than are red ones. (3) **Chromatic difference of spherical aberration** is a second order effect, consisting in the observation that the amount of spherical aberration for rays of different colors will be different in the same system. Red and blue rays (solid and dashed lines, respectively) incident to the system at the same distance from the axis (heights labeled I and II) are not brought to a focus at the same point. This effect becomes more pronounced in microscope objectives of wide aperture, leading to disturbing colored fringes around objects. Under certain illumination conditions and with certain techniques of correction for ordinary spherical and chromatic aberration, a yellowish coloring could taint part of the image. (4) **Chromatic difference of focal length or magnification** can remain, even when ordinary spherical aberration and its chromatic difference have been corrected. Rays of different colors and different incidence heights are brought to a focus at the same horizontal distance from the lens, but the rays of different colors no longer share the same degree of magnification. This also results in colored fringes, usually near the margin of the field of view.

parts with respect to the axis and according to the different refrangibility of different colors – [this magnification is] unequal when one compares the different partial images with one another, as also when one compares those within the same direction in the field-of-view.[62]

Fulfillment of the sine condition (together with the usual correction of spherical aberration) was sufficient to correct geometric errors for monochromatic light. With rays of many different colors, on the other hand, a number of

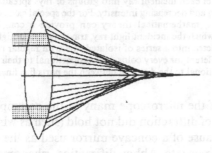

Spherical Aberration corrected
only for shaded zone

Figure 9. While ordinary spherical aberration could often be completely corrected in instruments with small angular aperture (such as astronomical telescopes), when apertures were wide, as was the case in all higher powered microscope objectives, often only a partial correction could be obtained. Abbe argued that many opticians corrected for a zone near the margins, making the objective's performance apparently better for viewing the tiniest of test objects with oblique illumination. In his view, this made the system unreliable for viewing anything else. Zeiss objectives were corrected for a zone somewhere between center and periphery.

other types of aberrations needed to be considered, some of which were difficult or impossible to correct completely. (See Figures 8 and 9.)

RESOLUTION AND RESEMBLANCE

These geometrical aberrations were not, however, the only limitations on the performance of the microscope to which Abbe devoted attention. It was well known that the quality of the image often seemed to be much more dependent on the size of the objective's aperture angle than on the degree of correction, especially with tiny diatoms and fine gratings. Dioptrical theories of image formation had been of little help in explaining this empirical relationship between aperture and image quality, and one could find numerous contradictory explanations in textbooks of microscopy. Abbe was able to add some experiments, however, which seemed to demonstrate that wider aperture improved image quality substantially only when the objects were extremely small; other experiments showed that when the object consisted of finely spaced lines or stripes, oblique illumination allowed resolution of finer details than direct illumination when the angular aperture of the objective was held constant. Furthermore, he claimed that it was the obliquity of illumination with respect *to the axis of the microscope* rather than with respect *to the plane of the object* that made the difference, and that this gave the final clue necessary to discover the true process at work in the resolution of very small objects:[63]

According to the wave theory of light, the phenomenon of diffraction is a characteristic alteration which material structures, according to the size of their dimensions, impose on light rays that pass through them, which are refracted by them or which are reflected by them. This modification consists in the dissolution of each incident ray into groups of rays spread out over a greater angle with periodically increasing and decreasing intensity. For the special cases of regular stratifications, stripes, rows of dots, etc., mathematical theory can provide a complete description of the phenomena, according to which the incident light ray, traveling in a straight line towards the other side [of the object] is diffracted into a series of isolated rays with regular angular separation; these angular separations are different for every color, and proportional to their wavelengths ... and are otherwise inversely proportional to the distance between the parts [i.e. lines, or dots] in the original structure.[64]

That in the case of the microscope many of the assumptions of the formal mathematical theory of diffraction did not hold (such as when the incident rays were not parallel because of a concave mirror used as the source of illumination) made no difference to Abbe; diffraction phenomena were at work. Diffraction allowed Abbe to explain away Nägeli and Schwendener's puzzling observation that the peripheral zones of wide-angled microscope objectives appeared to go unused; according to Abbe's theory, diffraction in the object created spectra that would be more widely spaced, the finer the detail being resolved, thus utilizing all of the objective's aperture.[65]

By way of demonstration Abbe described a set of experiments using various gratings as objects and involving the observation of the aperture image projected over the objective; the diffraction spectra brought about by the passage of the light through the gratings could be easily observed in this image by removing the ocular after the microscope had been properly adjusted and peering down into the tube (Figure 10, Column A). When the divisions on the grating were sufficiently fine (more than 30–50 lines/mm) and when the deployment of a stop which prevented all but the central maxima in the aperture image (Abbe said that this maxima consisted of the "un-diffracted" rays) from taking part in the formation of the image, no detail could be observed whatsoever; only an undifferentiated field of gray would be seen when the ocular was replaced (Figure 10, Column B). If, on the other hand, a different stop allowed the first diffraction maxima to pass through but excluded the central beam, an image of the object could be seen, but without detail (only equally bright, flat lines) and would appear against a dark background. Yet another stop, admitting two diffraction maxima, allowed some detail to appear, and the image became ever more detailed as more maxima were allowed through by different stops (Figure 10, Column C).[66]

The recombination of diffraction spectra allowed for the resolution of more detail. Since these spectra would be more widely spaced the finer the grating (and therefore the smaller the distance between lines), wider apertures allowed more spectra to pass through the optical system, and therefore resolved finer details. Oblique illumination resulted in the resolution of finer detail at the same aperture because higher order maxima were often allowed to enter the objective when the central beam was shifted all the way to one edge by the inclination of the illumination source with respect to the optical axis.

Abbe then constructed stops that would allow only non-consecutive diffraction maxima to be combined in the image; he found that clear images were

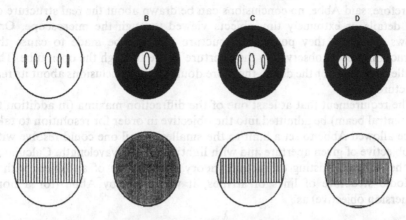

Figure 10. Abbe's experiments. These schematic illustrations show the effect on the image (bottom row) of selectively excluding certain diffraction maxima (top row) brought about by a grating. A (top) shows the normal spectra caused by a regular grating as viewed in the microscope tube with the ocular removed, and A (bottom) shows the resulting image. If a stop excludes all but the central maxima B (top), then only an undifferentiated, grey field remains in the image B (bottom). If the stop is widened somewhat C (top) so that the first of the secondary maxima now passes through, detail once again begins to appear C (bottom), but is somewhat coarser than in the original image. Adding and then widening the stop effectively limits, and then increases the angular aperture of the objective. If a stop with two holes is created so that only two non-sequential maxima are allowed to pass D (top), the image appears to have more lines, spaced more closely together. Abbe's article was not illustrated. These drawings are adapted from J.W. Stephenson, "Observations on Professor Abbe's experiments illustrating his theory of microscopic vision", *Monthly Microscopical Journal* 17 (1877), 82–8 and Dippel (1882).

formed that appeared to possess good detail, but that the separation between the divisions would appear halved, thirded, or quartered, and that the number of divisions would appear to be doubled, tripled, or quadrupled. With two or more systems of lines divided at crossed angles, by picking and choosing which maxima were admitted, Abbe could make a number of different patterns appear in the image.[67] These experiments, he argued, showed that images of very fine details do not necessarily resemble the objects they came from:

As these experiments make clear, those who refer to the familiar assumption of resemblance between an object and its optical image must now conclude that under these circumstances such an assumption is completely arbitrary . . . That different structures always yield the same microscopic image when the differences between the diffraction effects associated with them are artificially removed, and that the same structures always yield different images when their diffraction effects in the parts that are used by the microscope are made artificially different stands against this assumption. This is to say that the images of structure formed by the combination of diffraction spectra have no necessary relationship with the actual properties of the objects themselves, but only with the properties of the diffraction spectra which take part in the formation of the image.[68]

Therefore, said Abbe, no conclusions can be drawn about the real structure of fine details or extremely tiny objects viewed through the microscope. One knows only that they possess a structure that can be made to cause the diffraction spectra observed in the aperture image through the open tube. The smaller the object or the detail, the more doubtful the conclusions about its real structure.

The requirement that at least one of the diffraction maxima (in addition to the central beam) be admitted into the objective in order for resolution to take place allowed Abbe to set a limit to the smallest detail one could resolve with an objective of given aperture and with light of a given wavelength. Calculated on the basis of existing diffraction theory for the case of an object with a periodic structure of lines or stripes, it was given by Abbe (for a non-immersion objective) as

$$d = \frac{\lambda}{\sin\alpha}$$

with direct illumination, and by

$$d = \frac{\lambda}{2\sin\alpha}$$

with oblique illumination, where d is the smallest separation that can be resolved, λ is the wavelength of the light, and α is half the angular aperture of the objective. This limit, though reached using different means, was exactly equal to the limit published independently by Helmholtz just a few months later.[69]

Because the representation of structural detail depends on the combination of diffraction spectra, and because these spectra will be incident throughout the full range of the angular aperture of the objective, Abbe argued that it was absolutely crucial that geometric and chromatic errors, in particular those he identified as errors of magnification, be corrected. They could shift the size and position of diffraction spectra which entered different parts of the objective, and therefore might cause them to be inappropriately recombined just as they had been in the experiments where Abbe had manufactured false images by combining non-sequential diffraction maxima. Without correcting these errors, one also ran the much greater risk that the "diffraction image" of fine detail would not be properly superimposed on the normal "absorption" or "dioptrical image" of gross structure, formed by the familiar processes of geometrical optics:

The individual elements of the image in the microscope, the absorption image as well as the various component spectra of the diffraction image, are without exception brought about by isolated light-cones of relatively small divergence angle – almost never more than 30–40°. Even with a considerable amount of uncorrected spherical aberration, the tips of these isolated pencils, each considered singly, can be sharp enough to leave no noticeable circle of dispersion. But when a large aperture angle is employed these individual pencils make use of very different parts of the free aperture and therefore, in certain cases, may not be recombined at their [focal] tips; rather they will cross behind and next to one another. Then the component parts of the image will not come into correct combination, but rather will be displaced laterally and longitudinally. The structural landmarks that belong in the same place or at the same level in the object, such as systems of stripes, will therefore appear to be separate, and separate from the contours of the parts of the object to which they belong as well . . . I am certain that I would not be mistaken if I were to express the opinion that the consequences of these considerations probably play an unexpectedly great role in many of the controversies that rage between microscopists over the relationships between structures.[70]

The connection between these investigations and the advertisement for "scientifically" made objectives from Zeiss (with which Abbe's paper opened) now became explicit. "Because the first requirement that scientific use makes of the microscope is that what goes together in the object is represented as going together in the image, for every structure and regardless of the kind of illumination, it follows that equal correction of spherical aberration for the entire range of the aperture must be the guiding principle in the construction of the microscope" – and if one does not wish to be misled by what one sees, this requirement was even more important than wide aperture in a quality microscope. The objectives mentioned in the new 1872 catalog from Zeiss, Abbe reported, had all been constructed with this consideration in mind, and never was wideness of aperture pursued at the cost of the required off-axis correction.[71] The implications for the microscopist were clear: buy your microscope from an optician who doesn't understand these principles and you risk your reputation as an observer.

PLEASING THE CRITICS

Abbe's work leading up to the new Zeiss microscope objectives of 1872 and the publication of his paper on the theory of the microscope of 1873 have often been pointed to as demarcating the beginning of a new era in the relationship between theory and practice in optics.[72] Abbe, it is said, put the manufacture of microscopes (and in the process, of optical instruments more generally) on a "truly scientific" basis, at the same time establishing Zeiss' position as the leader in the field of precision optics. But at the time Abbe's work was made public, these consequences were not yet apparent, and it was not until much later – after considerable reinterpretation by microscopists, physicists, and Abbe himself – that his work was read in this way. During the 1870s and early 1880s, the meaning of Abbe's publications concerned the creation of demand for and defense of new products that he and Zeiss were introducing to the market, far more than it concerned the relationship between physical or geometrical optics and the technologies of microscope production. Optical theory as *publicly* employed by Abbe had far more to do with selling microscopes than with building them.

Zeiss's new objectives received very good reviews in the German scientific press. Leopold Dippel praised the new Zeiss objectives as among the best ever. He especially liked the degree to which Zeiss' new dry systems seemed to be free from aberration, providing exceptionally clear images, even though they lacked the superior resolving power provided by some competing systems:

Of all of the properties of these new dry systems that I have observed, there are a few I wish to emphasize. With direct illumination, the absence of color in organic preparations is as complete as I have seen. *Only* with strong ocular magnification and *only* in the weaker systems can a hint of yellow coloration be seen. Also the boundary lines of finer structures appear to be drawn exceedingly sharply and clearly, even in the weaker models. Finally, in the presentation of detail when resolution is tested they perform extremely well. Even if other systems of the same strength deliver quantitatively the same (or here and there a little more) [resolving power], I have never seen such beautiful images in any other objectives – with the exception of Winkel's IX – with respect to color, partial fogginess [*theilweiser Verschleierung*], or simultaneously sharp presentation of object boundaries and details (in the scales of diatoms, for example).[73]

Others shared in these opinions and the highest degree of correction of aberration, both on and off the optical axis became (and continued to be) the hallmark of the Zeiss objective systems. As Abbe had advocated in his article of 1873, even where they were widened, apertures were kept moderate.[74]

Even though sales at Zeiss were already growing, sales of microscope objectives significantly increased during the first few years after 1872. Much of this growth came from sales of the newer dry models designed by Abbe, and a substantial part was owing, no doubt, to the positive opinions of all of the Zeiss systems expressed by well-known microscopists. But little of this growth can be attributed to microscopists flocking to the one, true source for instruments constructed with "every curvature, every thickness, every opening" determined ahead of time on the basis of optical theory.

Some of the doubts that microscopists expressed were justified. Despite Abbe's claim that all of the parameters in his objective systems were fixed in advance, with only a single distance between lenses in the case of the strongest objectives left undetermined until the end of the fabrication process, those who had the opportunity to examine a number of different systems found considerable variation among different examples of the same type. Dippel had found a number of irregularities in the Zeiss objectives. While comparing seven examples of Zeiss's strongest dry system (F), he found that they seemed to possess quite different degrees of resolving power, and that one or two seemed to be notably inferior to the others. He returned them to Abbe for examination; Abbe wrote back with an explanation:

Concerning the results of your comparisons of the seven F systems which Mr. Zeiss had sent, the differences that you found can most simply be explained by my subsequent measurement of the aperture angles. Only #309 has the intended aperture of 105° (exactly.) The rest all have slightly smaller angles, the lowest, #311, with 96°. Since *Pleur[osigma] ang[ulatum]* already lies very near the limit of resolution for direct illumination for an objective with aperture angle of 105° – actually resolved only by the *most oblique* rays from the edge of the mirror – the difference of 9° turns out to be decisive. Small variations in the aperture angle – around 1–2° – are unavoidable in any case.[75]

The larger differences, Abbe said, appeared to be due to sloppy work on the part of one of Zeiss' craftsman, and he assured Dippel that the fabrication of these lenses would be much more tightly controlled in the future. Still, testing notebooks kept in the workshop during this period indicate that there was considerable variation among several parameters of the system (such as the thickness of cover-glass for which the lenses were corrected) and the overall quality of the corrections;[76] many years later, Abbe admitted to Dippel that the aperture angles advertised in the catalogs were only an approximation.[77]

Clearly, much continued to depend on the craftsmen, even when the microscopes were manufactured "according to theory".

The part of this theory published by Abbe also received some attention from
German microscopists. In particular, botanists who had written extensively on
the optics of the microscope and its relation to the new cell theories – men like
Dippel and Schwendener – read Abbe's work as highly relevant to their own
concerns about insuring the reliability of microscopic evidence.[78] They were
interested in what Abbe had to say about the limits of resolution and the role of
diffraction in the formation of images of fine structure, but they were even
more interested in what he had to say about the evaluation and testing of
instruments and the problems that accompany uncorrected errors of magnification. The objects with which these biologists typically dealt were much larger
than the tiny diatoms and fine rulings whose images Abbe claimed were entirely
caused by diffraction and interference effects, but the need to correct and test
for aberration in all zones of the objective to avoid false relationships between
structure and contours and between structural elements seemed to them to be
of immediate importance.

Dippel's earliest letters to Abbe concerning his theory of the microscope
questioned Abbe's calculation of the limit of resolution; Dippel claimed that
with a perfectly ordinary setup, he was able to resolve lines much more finely
ruled than Abbe seemed to think was possible. Abbe replied that his calculated
limit assumed "purely central illumination"; with an ordinary plane-mirror
reflecting light from the sky as a light source, the mirror would also provide
oblique rays which would assist in the resolution. Under these conditions, the
proper limit for a dry objective would be

$$d = \lambda \, / \, (\sin \alpha + \sin \beta) \text{ (rather than } d = \lambda \, / \, \sin \alpha)$$

allowing for the presence of the oblique rays from the edge of the mirror. Here
d is the smallest resolvable separation, λ is the wavelength of the light used, α is
the semi-aperture of the objective as before, and β is one half the angular
diameter of the mirror, measured from the intersection of the object and the
optical axis. Thus Abbe tried to show that Dippel's observations did not
contradict his findings at all. Abbe also gave a more general, experimental
justification for the basis of his calculation:

If you wish to test the fundamental principle behind my calculation of the limit, there is no easier
way than by observation of *Pleur. ang.* and similar objects with strongly delineated systems of
stripes. Place *Pleuro. ang.*, for example, under an immersion objective under direct illumination,
remove the ocular and bring the pupil of the eye to the location of the image brought about by the
objective. You will then see in the free aperture of the objective an image of the mirror surface
surrounded by 6 diffraction spectra, which, with the exception of their red ends on the outside, are
all completely visible. This is why the image contains so strongly the well known markings. Now if
you exchange that objective for a dry system whose aperture is less than 100°, you will see these
spectra obscured up to their blue ends on the inside; the image is now very difficult to make out. If
you then narrow the cone of incident light so that the angle designated above with b becomes
smaller – either by moving the mirror farther away or by covering its periphery, but better still by
using a cylindrical stop – then the 6 blue flecks will become weaker and weaker until they disappear
entirely behind the dark edge of the lens opening. *As soon as this point is reached, the last trace of the*

markings will disappear in the microscopic image [as viewed through the ocular] and one sees only an undifferentiated brown scale. Similar experiments can be performed with a Nobert grating. They are quite easily done with good, brightly lit clouds or with an instrument near a flame.[79]

For Dippel, however, the ultimate limit to the resolving power of the microscope was merely a curiosity. He was a botanist, and the objects in which he was most interested were nowhere near that limit. Therefore the subject of their correspondence quickly turned to the proper evaluation of microscopes and to a new method for testing them proposed in Abbe's paper. This test involved using a relatively coarse grating as a test object, and carefully regulated illumination using a wide-angled condenser that Abbe had specially developed for the purpose. Using the condenser, one created an illumination pencil whose angular aperture was greater than that of the objective, thus completely filling it with light. One then used a series of stops which allowed different parts of the illuminating beam (and therefore different zones of the objective's aperture) to take part in the formation of the image, checking the correction of Abbe's "errors of magnification." In particular, one could test directly for what Abbe claimed was the most damaging consequence of letting such errors persist: the improper blending of partial images formed by different zones at different distances from the optical axis.[80]

By the fall of 1873 Dippel was planning to undertake a comparison of objectives from different makers using Abbe's test. Abbe wrote to Dippel to say that he was happy to assist in any way he could, and was sending a new Zeiss microscope, a set of objectives, one of the illumination condensers described in the article, a set of gratings, and detailed instructions.[81]

An unstated purpose of this testing method was to give microscopists a concrete way of implementing the redefinition of performance that it embodied. Previously, microscopists had rated microscopes using several unconnected tests of various aspects of performance, the most important of them being resolving power. Textbooks of micrography often contained lists or tables of diatoms, ranked in terms of their difficulty of resolution, and a system's ability to resolve a difficult test object was usually taken as a good indication of its overall quality.[82] Abbe wanted to change all that. When an object was just barely resolved, he argued, only the zones of the objective at the center and the edge (where the central and first diffraction maxima enter) were used. The performance of the microscope under such conditions would therefore be quite different from its performance during normal observation of larger objects with coarser details, when all of the objective would come into play. A test of resolution *only* told one about the *limit* of resolution, and one could usually calculate that in advance; it never could give a means of comparing the overall quality of different objectives. In Abbe's view the best systems were not the ones that resolved the smallest objects, but the ones that provided the best all-around definition and which contained the highest degree of correction for differences between the various zones. These were the qualities for which the new Zeiss objectives were praised.

The role of Abbe's testing method in re-educating microscopists to value objectives with different qualities can be glimpsed in an exchange between

Abbe and Dippel concerning a dry objective with short focal length and very wide aperture constructed by a maker named Winkel. In correspondence with Abbe and in his earlier published reviews, Dippel had high praise for Winkel's objectives, and he believed the Winkel No. 9 system to be the best anywhere.[83] Abbe had argued that for dry systems, apertures over 110° or focal lengths below a few millimeters would lead to uncorrectable aberration and difficulties in merging the contour and structure images (formed by the separate dioptrical and diffraction processes).[84] Winkel's systems seemed to violate both of these limits. Abbe was counting on use of his test method to convince Dippel that Winkel's objective could not possibly be as good as he thought it was:

I am quite anxious to hear the results of your test of Winkel's No. 9. From what you have told me it must have a much wider aperture than our dry systems – 130–140° at least – and therefore, *if it really is well-corrected*, it must be a much better objective than our F. But so far, and I have tested a great many objectives, I have never found a single dry system of greater than 100° aperture angle that could stand the trial. Either oblique incident light-cones gave no sharp contours at all (with narrow but *pure* colored fringes) and instead gave a wide fuzzy border; or sharp definition could be had only with some arrangement other than central illumination. In the remarks on page 458–9[85] of my essay you will see that I believe that this type of error is the worst and most dangerous that a microscope system can possess.– According to theory, it *must* occur whenever the aperture angle exceeds the given measure, *to the extent that the surfaces are truly spherical.*[86]

The last possibility gave Abbe an out should the Winkel objective do well in the test after all. But Abbe was quick to point out that if the lens surfaces *were* in some way aspherical, they could not be manufactured with any kind of regularity, and every specimen of the objective would likely have very different properties. In short, according to Abbe, a system not constructed within the limits he specified would either be shown to deliver unreliable results, or to have been built using unreliable methods of fabrication.

This kind of argument often left Abbe in the unenviable position of having to show why some systems *seemed* superior even though they were not, at least according to his definition:

As I have already remarked, the greatest possible compensation for the critical difference of spherical aberration between red and blue rays should be sought so that in an objective of considerable aperture angle the point of most complete achromatism is not on the axis and not at the edge, but rather at a point in a zone between them – as is so of all of Zeiss' systems which with *moderately* oblique illumination show purely secondary colors in the middle of the field of view. If you find with Schieck and with Hartnack's No. IV that when you change to oblique light yellow and blue immediately become visible, then that will prove that the point of best achromatism is very near the axis – and therefore because of the small inclination of these rays, fringes do not become visible – or perhaps that the axis is already *over*corrected. In the first case the edge will be too overcorrected for the oblique rays to form a sharp image. If the opposite is true – as I have often seen in Merz's and, if I am not mistaken, also once in Hartnack's immersion systems – then the image of diatom skeletons with oblique illumination may be much more elegant than with my method of color compensation, but for direct light these lenses are almost completely unusable.[87]

Dippel's tests (using Abbe's methods and equipment) of Winkel's objective at first seemed to show it to possess both a wide aperture and good performance in combining direct and oblique test images. But Abbe was still unconvinced, and urged Dippel not to take the objective's performance at face value. If it worked well, he said, it *had to be* a fluke:

I have myself, when earlier I too took resolving power to be the measure of perfection and when Zeiss wanted to emphasize it and it alone, designed dry systems of 120–130° aperture angle with 1.5–2.5 mm focal length ... But after I was most decisively convinced that such objectives could not be made correctly and regularly, the effort to improve dry systems in this fashion was finally given up; and I will not consider this conclusion to have been a mistaken one until I have seen at least a half dozen such systems in a row delivered by the same craftsman with the same quality as *your* W[inkel No.] IX.[88]

Indeed, Abbe's testing method when combined with a suitable presentation of his work on the theory of the microscope seemed to be such a powerful tool for demonstrating the superiority of Zeiss' systems, that Abbe urged Dippel to discuss it in print in a forum likely to reach a large number of microscopists. Zeiss was prepared to undertake manufacture of the apparatus in whatever scale necessary, and Abbe envisioned a test-plate on the stage of every scientific user of the compound microscope:

Because your exacting scrutiny of the testing methods that we have been discussing and your methodical advances lead me to believe that you will also recognize it as useful and rational, I allow myself to suggest that you use the opportunity of your report planned for publication in *Flora* to discuss the simpler [testing] procedure with the familiar concave mirror. I think – completely aside from any personal interest on the part of Zeiss or myself – that it would certainly be to the advantage of microscopical studies if a stricter measure for the evaluation of objectives were to come into general use and the manipulations required adapted so far as possible to the [majority of] microscopists; then the force of competition could not help but increase the average quality of microscopes more and more.[89]

Zeiss was making test plates, said Abbe, in sufficient quantity that they could be included free of charge with every microscope that Zeiss delivered. Dippel obliged in an article that appeared towards the end of 1873, and described Abbe's test method both with and without the special condenser, and included a notice that the test plates could be had from Carl Zeiss.[90]

Dippel was evidently swayed by Abbe's arguments concerning the Winkel objective as well. The same article contained the results of his comparisons of various objectives according to the new test methods, and in it Dippel moderated his praise of the Winkel systems, saying that "only in single systems" was the correction of spherical and chromatic aberration superior to that of Zeiss. Though some objectives from other makers might have wider apertures and greater resolving power than Zeiss', "one need only consider quality more exactly in order to abandon the prejudice in favor of larger aperture and the higher resolving power that it affords".[91]

Other biologists who had written on the optics and theory of the microscope also took an active interest in Abbe's work. Simon Schwendener found Abbe's work so persuasive that he thought it justified a complete revision of several large sections of his book.[92] The second edition of Nägeli and Schwendener's *Das Mikroskop* appeared in 1877, and contained extensive discussion both of Abbe's diffraction theory and his recommendations on the testing and evaluation of objectives.

Nägeli and Schwendener strongly endorsed Abbe's explanation of the function of aperture and his account of the formation of images of fine structure by the refocusing of diffraction spectra brought about when light passes through tiny objects:

As Abbe has shown, there is a second way of much greater importance and higher significance in which the size of the angular aperture exerts a mathematically provable and experimentally confirmed effect. It has nothing to do with correction of coincidental imperfections of construction, but rather with a specific function of the aperture angle with respect to rays that are deflected in the object plane and which interfere in (or near) the upper focal plane. It can be shown that it is by means of this function alone that the fine structural details of the object, such as the stripes of diatoms, are made perceptible in the microscopic image. . . .

The importance of the aperture angle can therefore be found in the fact that it makes possible the formation of the interference image, in which all of the finer details are contained.[93]

The two also gave a great deal of attention to Abbe's pronouncements on the importance of the correction of aberrations for all zones of the aperture to insure a proper merging of the dioptrical and diffraction images:

One is used to rating the performance of an objective according to its resolving power with oblique illumination; therefore the efforts of opticians have been directed at increasing the aperture angle as much as possible and, so to speak, to accommodate the entire design of the objective to the resolution of particular test-objects. The single-mindedness of such procedures can be seen immediately upon closer consideration. The limit of the defining power for a given objective is clearly found in such details from which, aside from the direct rays, the first diffracted light-pencil barely enters the objective. The two light-cones that are used strike the objective therefore in two diametrically opposite points on the edge of the objective, the direct rays on the side opposite the mirror, the diffracted rays on the same side from which the light comes; the entire middle part of the aperture does not even come into consideration. As long as the objective does not show any adverse aberration for this narrow edge-zone – which as we have said can always be obtained, even for poor construction, simply by changing the distance between the lenses – then the two images will necessarily be completely superimposed; the performance of the objective will be satisfactory in this special case, and may even seem to be quite good. But it tells us nothing more than that the aperture is large enough to allow at least of the light pencils that are necessary for the formation of the image, one diffracted and one undiffracted, to enter the objective and that its design at least allows to possibility of correction for a narrow peripheral zone. The observer therefore learns in this way only the *limit* of definition, not the ability of the objective to define [objects] in more usual cases of practice.[94]

Nägeli and Schwendener agreed with Abbe that it was necessary to redefine "good optical performance" in terms of new testing procedures and they repeatedly emphasized that it was not enough to see whether or not an object will resolve fine gratings or tiny diatoms if one wants to judge how an objective will perform under normal conditions. Like Dippel, Nägeli and Schwendener advocated the adoption of Abbe's methods of testing several zones of the aperture against one another.[95]

It is difficult to say to what degree Abbe's methods were actually adopted by the "rank and file" of practicing microscopists in Germany; certainly not every microscopist of the mid-1870s embraced Abbe's theories as strongly as did Dippel, Nägeli, and Schwendener – influential though they were. Commentary on Abbe's work occurred primarily in occasional review articles evaluating the latest microscopic apparatus and in sections of microscopy textbooks that dealt with the optics of the microscope or with procedures for evaluating their performance. But sales at Zeiss continued to climb, and the company was gaining a reputation as the leading German source of microscopic apparatus.

Abbe shared in the profits in several different ways. In 1870 he had been promoted to extraordinary professor, largely on the basis of research he had published or promised dealing with his work on the microscope, and in December of 1873 he was admitted to the German Academy of Natural

Science.[96] In 1875, he asked Zeiss to be allowed to participate materially in the optical business. Zeiss made him a two-fifths silent partner, and gave him a third of the profits.[97] Abbe was now both a professor and a businessman.

EPILOGUE

In 1870, the microscope was just beginning to attract attention from physicists willing to consider the particularities of microscopical optics. Helmholtz's article was read by physicists as an indication that the issues affecting the performance of the microscope were in many cases quite different from those affecting optical instruments with which they were more familiar, such as the astronomical telescope.

But Helmholtz' work did little to affect either the construction or the use of the instrument. It was the practice of experiment and measurement, rather than the theory of physical or geometric optics, that proved to be the more important contribution from physics where the design and fabrication of microscopes were concerned. Though Abbe came to Zeiss with nothing more than basic training in the physics of light, the practical techniques of precise measurement he applied to the tiny dimensions and particular needs of the microscope workshop made an almost immediate impact on Zeiss' ability to fashion quality optical systems. It also made it possible to begin to introduce the theoretical tools.

But it was not always easy to convince those who bought and used microscopes that what Abbe *thought* was an improvement really was, and his efforts to publish and win acceptance for his ideas on the formation of images in the microscope should be read in this context. Appearing primarily in journals aimed at biologists, not physicists, they were first and foremost efforts to reeducate users of the microscope so that they could appreciate the features that made his objectives, produced by Zeiss, superior. There was no widespread agreement among microscopists and it was not always obvious which systems did a better job – much depended on the particular circumstances under which they were used – and microscopists advanced and used theories to link contrived testing procedures to observational outcomes in more normal conditions. Abbe's paper was read by German microscopists in the early 1870s as being primarily about these theories.

Debates over the theory of microscope testing and evaluation could be extremely intense, and when new technologies appeared to produce results that were inconsistent with the accepted means of testing and evaluation, microscopists would sometimes refuse to buy the new apparatus before they bought the theory which explained why it was superior. But when the battle over the initial acceptance of Abbe's lenses was won, and microscopists were convinced of the superiority of Abbe's testing method, elaborating on the diffraction mechnanism in the theory of the microscope had become a very low priority. Abbe never did publish the more rigorous treatment he had promised, and most of the technical, geometrical optics he had done related to the theory of instruments remained the property of the Zeissworks, handed down through university-like seminars held at the plant.

In 1891, at the *Naturforscherversammlung* in Halle, Abbe gave a reprise demonstration of some of his experiments on the role of diffraction in the formation of microscopic images. Sigfried Czapski, Abbe's assistant and designated successor at Zeiss, was shocked to discover the surprise with which the assembled physicists greeted Abbe's nearly 20 year old demonstrations. Few of them knew of Abbe's work on the theory of microscopic images at all, and most of those who did had assumed that Abbe had little to add to what Helmholtz had said in his 1874 article in the *Annalen der Physik*. Virtually none of them was familiar with his publications in the microscopical literature; among physicists Abbe's reputation was as an instrument maker, not a theoretician.[98]

Czapski was already well aware of the confusion that existed over the differences between Abbe's and Helmholtz's contributions to the theory of the microscope – in particular of the widespread perception that the issue of a theoretical limit to the resolving power of the microscope had not moved beyond the point where Helmholtz had left it in 1874. This confusion had been the subject of a short piece by Czapski explaining the key difference between the theories of Helmholtz and Abbe (Helmholtz treated the case of self-luminous objects, while Abbe treated non-self-luminous ones) and arguing that it had been the Abbe theory – not Helmholtz's – that led to the definition of numerical aperture and inspired the use of high-index immersion fluids.[99]

After the Halle episode, Abbe dusted off his old lecture notes and returned to the classroom with an advanced seminar on the theory of diffraction in optical instruments, hoping to fill out the theory in its full mathematical rigor. Finally, he hoped to complete the thorough physical and mathematical treatment of the subject he had promised in 1873, but had never delivered. Though Abbe managed to generate some 255 pages of manuscript, the press of business and his own poor health ultimately prevented him from finishing and publishing it.[100]

At about this time, Czapski also was in the process of collecting and collating material on Abbe's considerable body of work on the geometric optics of optical instruments, accumulated over nearly three decades of efforts to develop and improve the products of the Zeiss company, and passed on to his followers through lectures at the University of Jena. Czapski published the material in a physical handbook edited by a physics professor at Jena, and then as an advanced textbook of geometrical optics under the title *The theory of optical instruments according to Abbe*, the first part of which appeared in 1891. Several sections of the book followed Czapski's notes on Abbe's seminars almost equation for equation. This volume became an important text for physicists who worked with or designed optical instruments, and its continuous extension and revision became a house project at the Zeissworks. Virtually all of the top-level *wissenschaftliche Mitarbeiter* at Zeiss contributed sections to later editions of the book.[101]

Finally, amid preparations for Abbe's impending retirement, word reached the Zeissworks in 1903 that Abbe had been nominated for a Nobel Prize. Fully aware of the prestige that such an award would bring to Abbe and those associated with him, and hoping that it might increase awareness of Abbe's achievements among those in a position to influence the award, the Zeiss *Geschäftsleitung* immediately authorized a crash program for issuing a series of Abbe's *Collected Works*. Funds were allocated and responsibility delegated to collect, edit and publish Abbe's scientific papers as soon as possible, beginning with his work on the theory of the microscope. Czapski began negotiating with a publisher right away, and the first volume appeared only a few months later.[102]

Abbe did not get the Nobel Prize, however. The physics prize for 1904 went instead to J.W. Strutt, the third Baron Rayleigh, for his work on gas theory and the discovery of argon. Ironically, Rayleigh had also published a contribution to the theory of the microscope in 1896, in which he had expressed surprise that Abbe had anything interesting to say on the matter:

It would seem that the present subject, like many others, has suffered from overspecialization, much that is familiar to the microscopist being almost unknown to physicists, and *vice versa*. For myself I must confess that it is only recently, in consequence of a discussion ... in the *English Mechanic* that I have become acquainted with the distinguishing features of Prof. Abbe's work.[103]

Like other physicists, Rayleigh had completely passed over Abbe's contributions to the theory of the microscope. Throughout his career, Abbe had always given priority to publications aimed at microscopists – the buyers of his products and the foundation of his firm – frequently promising, but rarely delivering, a more thorough treatment for mathematicians and physicists.

Deloitte & Touche Consulting Group, Two World Financial Center, New York, NY 10081-1420, USA

NOTES

I thank Jed Buchwald, David Cahan, Loren Butler Feffer, John Heilbron, Edith Hellmuth, and Joachim Wittig for comments and guidance. This material is based upon work supported under a graduate fellowship and travel grant from the National Science Foundation.

Abbreviations:
BACZ: Betriebsarchiv Carl Zeiss Jena
HStA: Thüringisches Hauptstaatsarchiv Weimar, Bestandsbezeichnung Carl Zeiss Jena

[1] For more on electrical technology and especially telegraph cables, see C. Smith and M. Norton Wise, *Energy and Empire: A biographical study of Lord Kelvin* (Cambridge: Cambridge University Press, 1989); D.S.L. Cardwell, *James Joule: A biography* (Manchester: Manchester University Press, 1989); B.J. Hunt, "Michael Faraday, Cable telegraphy and the rise of field theory", *History of Technology* 13 (1991), 1–19. For more on the relationship between 19th century science and technology, especially in the German context, see A. Beyerchen, "On the stimulation of excellence in Wilhelmian science", in J.R. Dukes and J. Remak, *Another Germany: A reconsideration of the*

imperial era (Boulder: Westview, 1988), 139–68; D. Cahan, *An institute for an empire: The Physikalisch-Technische Reichsanstalt, 1871–1918* (Cambridge: Cambridge University Press, 1989); J.A. Johnson, *The Kaiser's chemists: Science and modernization in imperial Germany* (Chapel Hill: University of North Carolina Press, 1990); R. Vierhaus and B. vom Brocke, eds., *Forschung im Spannungsfeld von Politik und Gesellschaft: Geschichte und Struktur der Kaiser-Wilhelm-/Max-Planck-Gesellschaft* (Stuttgart: Deutsche Verlags-Anstalt, 1990).

[2] See Stuart M. Feffer, "Microscopes to Munitions: Ernst Abbe, Carl Zeiss, and the transformation of technical optics, 1850–1914", Ph.D. dissertation, University of California, 1994, chapter 5.

[3] Feffer (1994), pp. 32–43.

[4] For accounts of early attempts to achromatize the microscope, see H. de Martin and W. de Martin, *Vier Jahrhunderte Mikroskope* (Wiener Neustadt: Weilburg, 1983), 24; S. Bradbury, *The evolution of the microscope* (Oxford: Pergamon, 1967), esp. 164–171; G. L'E. Turner, "Micrographia Historica: The study of the history of the microscope", in *Essays on the history of the microscope* (Oxford, 1980), 1–29, esp. 12; G. L'E. Turner, "The microscope as a technical frontier in science", *ibid.*, 159–183, esp. 162.

[5] On Fraunhofer, see M. v. Rohr, *Joseph Fraunhofers Leben, Leistungen und Wirksamkeit* (Leipzig: Akademie Verlagsgesellschaft: 1929); G.D. Roth, *Joseph von Fraunhofer: Handwerker-Forscher-Akademiemitglied* (Stuttgart: Wissenschaftliche Verlagsgesellschaft, 1976); M. Jackson, "Artisanal knowledge and experimental natural philosophers: The British response to Joseph Fraunhofer and the Bavarian usurpation of their optical empire", *Studies in History and Philosophy of Science* 25 (1994): 549–576. For one microscopist's evaluation of Fraunhofer's instruments, see H. v. Mohl, *Mikrographie, oder Anleitung zur Kenntniss und zum Gebrauche des Mikroskops* (Tübingen: L.F. Fues, 1846), 66, 71–5, 337.

[6] Bradbury (1967), 180–1, 184–5; C. Chevalier, *Des microscopes et de leur usage* (Paris: Proux, 1839); J. Fresnel, "Rapport sur le Microscope achromatique de M. Selligue", *Annales des sciences naturelles* 3 (1824): 345–54.

[7] J.J. Lister, "On some properties in achromatic object-glasses applicable to the improvement of the microscope", *Philosophical transactions of the Royal Society of London* 120 (1830): 187–200; Turner, "Micrographia Historica" (1980), 13–14.

[8] J. Lister, "Obituary notice of the late Joseph Jackson Lister, FRS, ZS, with special reference to his labours in the improvement of the achromatic microscope", *Monthly Microscopical Journal* 3 (1870): 134–43; G. L'E. Turner, *God bless the microscope: A history of the Royal Microscopical Society over 150 years* (Oxford: Royal Microscopical Society, 1989), 9–13.

[9] On the training of microscope makers, see Feffer (1994), 19–21.

[10] J. Wittig, "Friedrich Körner und die Anfänge des wissenschaftlichen Gerätebaues in Jena in der ersten hälfte des 19. Jahrhunderts", *NTM* 18 (1981): 17—28; H. Koch, "Die jenaischen Universitäts-Mechanici", Zeiss-Werkzeitung 11 (1936): 34–7; H. Koch, "Johann Friedrich Braunau: der Amtsvorgänger von Carl Zeiss als Universitätsmechaniker", *Zeiss-Werkzeitung* 18 (1945): 9-10. Several other universities had similar arrangements with instrument makers.

[11] H.A. Willam, *Carl Zeiss, 1816–1888: Sechstes Beiheft der Tradition: Zeitschrift für Firmengeschichte und Unternehmerbiographie* (Munich: F. Bruckmann, 1967), 15–18.

[12] Willam (1967), 19–21.

[13] P. Blume, *Betrachtungen zur Entwicklung der Technik und Technologie des Feinmechanisch-optischen Gerätebaues bei Carl Zeiss Jena im Zeitraum 1846–1945*, Geschictskommission der IKL des SED des VEB Carl Zeiss Jena, BACZ, no number, no date, 39; "Manual I & II", HStA Weimar, I/3/1736.

[14] M. v. Rohr, "Zur Geschichte der Zeissischen Werkstätte bis zum Tode Ernst Abbes", *Forschungen zur Geschichte der Optik: Beilage zur Zeitschrift für Instrumentenkunde* 2 (1936–38): 1–119, esp. 11–12.

[15] Willam (1967), 58.

[16] H.C. Freifsleben, "Seidel, Phillip Ludwig von", in C. Gillispie, ed., *Dictionary of Scientific Biography* (New York: Scribners, 1970-1978), vol. XII, 289–90; C. Jungnickl and R. McCorrmach, *Intellectual mastery of nature: Theoretical physics from Ohm to Einstein, vol. I, The torch of mathematics, 1800–1870* (Chicago: Univ of Chicago, 1986), 275–8; M. v. Rohr "Die Voigtlandérsche optische Werkstätte und ihre Umwelt", *Zeitschrift für Instrumentenkunde* 45 (1925), 480. See also the reprints appended to A. Steinheil and E. Voigt, *Handbuch der angewandten Optik* (Leipzig: B.G. Teubner, 1891) and the historical notes in M. v. Rohr, ed., *Die Bilderzeugung in optischen Instrumenten vom Standpunkte der geometrischen Optik* (Berlin: Julius Springer, 1904).

[17] Erményi, *Dr. Josef Petzvals Leben und Verdienste* (Halle: Knapp, 1903), 1–25; J. Petzval, *Bericht über die Ergebnisse einiger dioptrischer Untersuchungen* (Pesth: Hartleben, 1843), *passim*.

[18] H. Harting, *Zur Geschichte der Familie Voigtländer, ihrer Werkstätten und ihrer Mitarbeiter* (Braunschweig: Voigtländer & Söhne AG, 1925), 25–36; M. v. Rohr (1925), 441–5.

[19] Edmund Hartnack, for instance, took on Polish mathematician Adam Prazmowski in 1864. See J.C. Poggendorf, *Biographisch-literarisches Handwörterbuch der Geschichte der exacten Wissenschaften* (Leipzig, 1863), vol. II, 522; B.W. Feddersen and A.J. von Oettingen, *J.C. Poggendorf's biographisch-literarisches Handwörterbuch der Geschichte der exacten Wissenschaften* (Leipzig: J.A. Barth, 1898), 3rd edition, vol. II, 1065; H. Weil, "Hartnack-Kurzbiographie", *Sammelinfo 2*, ed. by H. Baden, no date, in BACZ, no accession number. Zeiss attempted to manufacture the designs of Weimar mathematician Friedrich Wilhelm Barfuss, but these efforts apparently were unsuccessful. See F.W. Barfuss, *Optik, Catoptrik und Dioptrik* (Weimar: Fr. Voigt, 1839); F.W. Barfuss, "Ueber das Mikroskop", *Astronomische Nachrichten* 20 (1843): 17–24, 40–48; F.A. Nobert, "Ueber die Prüfung und Vollkommenheit unserer jetzigen Mikroskope", *Annalen der Physik* 67 (1846): 173–85; F.W. Barfuss, "Ueber die Construction zusammengesetzter Microscope", *Annalen der Physik* 68 (1846): 88–91; Willam (1967), 17; Wittig (1981), 24. Barfuss' book also went through a second edition: H. Gieswald, *F.W. Barfuss's populäres Lehrbuch der Optik, Katoptrik und Dioptrik* (Weimar: B.F. Voigt, 1860).

[20] The physicists who published about geometrical optics had their major orientation toward astronomical instrumentation. Of 22 authors of papers on the theory of optical instruments, 18 held university or other academic positions. Of these, eight had held positions with an observatory or as a professor of astronomy at some point in their careers, and 11 published articles on astronomical observations, aspects of theoretical astronomy, or geodesy. None was a professor in botany, anatomy, physiology, medicine or any other field in which microscope use was common. Only one ever published a paper describing any results from the use of a microscope, and this was a piece dating from the 1870s (by G.G. Stokes) on an investigation of crystals. For bibliography I am relying here on the "Autorenregister" and "Historische Notizen" from v. Rohr, (1904) and the several editions of *Poggendorf's biographisch-literarisches Wörterbuch* for information on the careers of the authors. Even though the collection might be regarded as arbitrary or selective, the point remains: most of those with mathematical training and an interest in geometrical optics tended to come from backgrounds in astronomy, not biology, and we should not be surprised that their orientation was heavily skewed toward astronomical instrumentation.

[21] M.J. Schleiden, *Grundzüge der wissenschaftlichen Botanik nebst einer methodologischen Einleitung als Anleitung zum Studium der Pflanze, Erster Theil* (Leipzig: W. Engelmann, 1842), 144–5.

[22] T. Lenoir, *The strategy of life: Teleology and mechanics in nineteenth-century German biology* (Chicago: Univ of Chicago, 1982), 112–155.

[23] Turner, "The microscope as a technical frontier" (1967), 161–77.

[24] E. Cittadino, *Nature as the laboratory: Darwinian plant ecology in the German Empire, 1880–1900* (Cambridge: Cambridge Univ., 1990), 14–16; J. Maienschein, "Cell theory and development", in R.C. Olby, G.N. Cantor, J.R.R. Christie, and M.S.J. Hodge, *Companion to the history of modern science* (London: Routledge, 1989), 357–73; Bradbury (1967), 201–2.

[25] Several infamous episodes are mentioned in Schleiden (1842), 137–9 and P. Harting, *Das Mikroskop: Theorie, Gebrauch, Geschichte und gegenwärtiger zustand desselben*, translated by F.W. Theile, (Braunschweig: Vieweg, 1859), 332–3.

[26] Cittadino (1990), 18.

[27] Schleiden (1842), 112–4.

[28] Schleiden (1842), 136–7.

[29] R. Virchow, "Ueber die Reform der pathologischen und therapeutischen Anschauungen durch die mikroskopischen Untersuchungen", Virchow's *Archiv für pathologische Anatomie und Physiologie und für klinische Medicin* 1 (1847): 207–255, on 207–8. See also R. Virchow, "Cellular-Pathologie", Virchow's *Archiv für pathologische Anatomie und Physiologie und für klinische Medicin* 8 (1855): 3-39, esp. 3–9.

[30] Another line of micrographic texts can be found descending from Theodor Schwann, another early cell theorist. See E. Kaiser, "Ueber die Entwicklung und gegenwärtige Stellung der Mikroskopie in Deutschland", *Zeitschrift für Mikroskopie* 1 (1877–78): 1–9, 33–41, 97–111, 161–75, 225–36, 257–72.

[31] B. Bracegirdle, *A history of microtechnique* (Ithaca: Cornell, 1978), *passim*.

[32] Harting (1859), 79–83

[33] They usually translated Goring's "penetrating power" as "*penetrierende*" or "*resolvierende Kraft*". Harting specifically cited both Goring and Herschel in introducing the vocabulary in his book, using "*Definierende Kraft*" or "*Begrenzungsvermögen*" for Goring's "defining power", but preferring the term "*Unterscheidungsvermögen*" for "penetrating power".

[34] Harting (1859), 250.

[35] L. Dippel, *Das Mikroskop und seine Anwendung* (Braunschweig: Vieweg, 1867), 81–2.

[36] Nägeli and Schwendener (1867), 80–81.

[37] F. Auerbach, *Ernst Abbe: Sein Leben, sein Wirken, seine Persönlichkeit nach den Quellen und aus eigener Erfahrung geschildert* (Leipzig: Akademische Verlagsgesellschaft, 1918), 52–89. On the special character of Weber's experimental program, see J. Buchwald, *The Creation of Scientific Effects* (Chicago: University of Chicago Press, 1994), 19–20.

[38] On Weber's seminar and its relation to physics pedagogy in Germany, see K. Olesko, *Physics as a calling: Discipline and practice in the Königsberg seminar for physics* (Ithica: Cornell, 1991), esp. 409–10. On Abbe's own experiences at Göttingen, see E. Abbe to C. Martin, 26 Dec 1859 and Abbe to Martin, 11 Feb 1860, in E. Abbe, *Briefe an seine Jugend- und Studienfreunde Carl Martin und Harald Schütz, 1858-1865*, V. Wahl and J. Wittig, eds., (Berlin: Akademie-Verlag, 1986), 9–15.

[39] On Meyerstein and Abbe's friend, see Abbe (1986), xvii; Auerbach (1918), 91. On Abbe and Meyerstein, see Abbe to H. Schütz, 13 Apr 1861, Abbe (1986), 69–70; Abbe to Schütz, 23 Jun 1861, Abbe (1986), 94–8; Abbe to Schütz, 4 Sept 1861, Abbe (1986), 119–22. For Abbe's lack of interest in secondary education careers, see Abbe to Martin, 25 Jul 1860, Abbe (1986), 30–33. For Abbe's *Habilitation* see E. Abbe, "Ueber die Gesetzmässigkeit in der Vertheilung der Fehler bei Beobachtungsreihen", *Gesammelte Abhandlungen*, vol. II, 55–81. When Abbe began teaching physics at Jena as a young *Privatdozent*, he told a friend that he felt much more at home with the theory of measuring instruments than he did teaching even basic mechanics; see Abbe to Schütz, 12 Dec 1863, Abbe (1986), 267–75; Auerbach (1918), 122.

[40] For Abbe and Listing's lectures, see Abbe to Martin, 11 Feb 1860, Abbe (1986), 12–5. For Abbe and the Göttingen observatory, see Abbe to Schütz, 23/25 Apr 1861 and Abbe to Martin, 6 May 1861, Abbe (1986), 71–6.

[41] J.B. Listing, "Vorschlag zu ferner Vervollkommung des Mikroskops auf einem abgeänderten dioptrischen Wege", *Annalen der Physik* 136 (1869): 467–72; J.B. Listing, "Nachtrag betreffend die neue Construction des Mikroskops", *Annalen der Physik* 136 (1869): 473–9; J.B. Listing, "Notiz über ein neues Mikroskop von R. Winkel", *Annalen der Physik* 142 (1871): 479–80.

[42] H. Helmholtz, "Die theoretische Grenze für die Leistungsfähigkeit der Mikroskope", *Annalen der Physik Jubelband* (1874): 557–584, esp. 561–570.

[43] Helmholtz (1874), 579–584.

[44] Helmholtz (1874), 584. Abbe's article will be discussed in greater detail below.

[45] Abbe (1986), xxvii.

[46] E. Abbe, "Ueber mikrometrische Messung mittelst optischer Bilder", *Sitzungsberichte der Jenaischen Gesellschaft für Medicin und Naturwissenschaft* (1878): 11–17, in *Gesammelte Abhandlungen*, vol. I, 165–72; S. Czapski, "Mittheilungen aus der optischen Werkstätte von Carl Zeiss in Jena", *Zeitschrift für Instrumentenkunde* 12 (1892): 185–97, esp. 185–89; v. Rohr (140), 28. For a brief discussion of contemporary methods of measuring the focal length of microscope objectives, see Nägeli and Schwendener (1867), 170–6.

[47] Czapski (1892), 186–91, 196–7.

[48] E. Abbe, "Neue Apparate zur Bestimmung des Brechungs- und Zerstreuungsverögens fester und flüssiger Körper", *Jenaischen Zeitschrift für Naturwissenschaft* 8 (1874): 96–174, in *Gesammelte Abhandlungen*, vol. II, 82–164, on 82–6.

[49] M. v. Rohr, *Ernst Abbe* (Jena: Gustav Fischer, 1940), 33.

[50] C. Pulfrich, "Ueber einige von Prof. Abbe konstruierte Messapparate für Physiker", *Zeitschrift für Instrumentenkunde* 12 (1892): 307–15.

[51] The method, though independently developed by Löber, has also been attributed to Fraunhofer. See Auerbach (1918), 213; M. v. Rohr, "Ueber die Arbeitgemeinschaft von Carl Zeiss und Ernst Abbe bis zum Ende der siebziger Jahre, I", *Forschungen zur Geschichte der Optik (Beilage zur Zeitschrift für Instrumentenkunde)* 2 (1936–38): 160–76, esp. 161.

[52] Many years later, Abbe described these attempts as a "year-long fiasco" [*jahrelange Misserfolge*]. See E. Abbe, "Nachruf auf Carl Zeiss", *Gesammelte Abhandlungen*, vol. II, 339–41, on 341.

[53] E. Abbe, "Über die Grundlagen der Lohnregelung in der Optischen Werkstätte", *Gesammelte Abhandlungen*, vol. III, 119–56, esp. 138; M. v. Rohr, "Über den Ausgang der Arbeitgemeinschaft von Carl Zeiss und Ernst Abbe, II", *Forschungen zur Geschichte der Optik (Beilage der Zeitschrift für Instrumentenkunde)* 2 (1936–38): 253–92, on 270; v. Rohr (1940), 34–40.

[54] Carl Zeiss, "Nr. 19. Mikroskope und Nebenapparate von Carl Zeiss in Jena, 1872", BACZ 30528; E. Abbe, "Beiträge zur Theorie des Mikroskops und der mikroskopische Wahrnehmung", Max Schultze's *Archiv für mikroskopische Anatomie* 9 (1873): 413–68, in *Gesammelte Abhandlungen*, vol. I, 45–100. For date of submission of the "Beiträge", see M. v. Rohr, *Ernst Abbe* (Jena: Gustav Fischer, 1940), 217.

[55] Carl Zeiss, "Nr. 19. Mikroskope und Nebenapparate von Carl Zeiss in Jena, 1872", BACZ 30528; Carl Zeiss, "Zusammengesetzte Mikroskope", August 1861, BACZ 30528.

[56] Only one objective had been given a major modification (it no longer had a removable middle lens). The objectives with wider aperture, while of a new design for Zeiss, were based closely on their narrower counterparts – in one case sharing the same front lens – and their apertures were still quite moderate when compared to current offerings from other makers. Zeiss' three immersion objectives were completely new (Zeiss had never sold immersion systems before) and their design was based on the wider-angled dry objectives, with an additional doublet added at the back to allow for correction of spherical aberration left uncompensated by the immersed front lens. E. Abbe, Notebook containing sketches, fabrication plans, and test results for microscope objectives, 1873–1879, BACZ 12443; H. Boegehold, "Zur Geschichte der Zeissischen Mikroobjektive bis 1940", *Jenarer Jahrbuch* (1951): 7–21.

[57] Carl Zeiss, "Nr. 19. Mikroskope und Nebenapparate von Carl Zeiss in Jena, 1872", BACZ 30528.

[58] Even into the 1890s, most physicists only knew of Abbe's work indirectly, through the note appended to Helmholtz's 1874 piece on the microscope in Poggendorf's *Annalen*; on the basis of those brief remarks, many assumed that the work of Abbe and Helmholtz on the theory of the microscope was largely interchangeable. Zeiss' catalogs began explicitly referring to Abbe's article in 1878. See Carl Zeiss, "No. 23. Illustrierter Katalog über Mikroskope und Nebenapparate aus der optischen Werkstätte von Carl Zeiss in Jena", 1878, BACZ 30528.

[59] Abbe, "Beiträge" (1873), 45–7.

[60] Abbe, "Beiträge" (1873), 47.

[61] Dippel, for instance, in his 1867 treatise on the microscope, repeatedly mentioned Zeiss' work as being among the best that Germany had to offer and in 1869 wrote that his products compared favorably even with those of Hartnack. See, for instance, Dippel (1867), 184; L. Dippel, "Mikrographische Mitteilungen", *Archiv für mikroskopische Anatomie* 5 (1869): 281–94.

[62] Abbe, "Beiträge" (1873), 56.

[63] In the article, Abbe merely summarizes the results of these experiments. He does not describe them in any detail. See Abbe, "Beiträge" (1873), 73–4.

[64] Abbe, "Beiträge" (1873), 75.

[65] Investigating Nägeli and Schwendener's claims about the unused "dark space" may have led Abbe to these experiments in the first place. See E. Abbe, "Ueber die Grenzen der geometrischen Optik mit Vorbemerkungen über die Abhandlung 'Zur Theorie der Bilderzeugung' von Dr. R. Altmann", *Sitzungsberichte der Jenaischen Gesellschft für Medizin und Naturwissenschaft* (1880): 71–109, esp. 75–80; an abridged version can be found in Abbe's *Gesammelte Abhandlungen*, vol. I, 273–312. See also J. Wittig, *Ernst Abbe* (Leipzig: G. Teubner, 1989), 63–4.

[66] Abbe, "Beiträge" (1873), 77–8.

[67] Abbe, "Beiträge" (1873), 78–80.

[68] Abbe, "Beiträge" (1873), 82.

[69] Helmholtz (1874), 557–84. In Helmholtz's treatment, the objects are self-luminous and diffraction is caused when the light pencil passes through the front opening of the objective. In Abbe's treatment, the objects are not self-luminous and the diffraction occurs in the object itself – an integral part of rather than an impediment to the process of image formation.

[70] Abbe, "Beiträge" (1873), 90.

[71] Abbe, "Beiträge" (1873), 90–1. The trade-off between off-axis correction and aperture was essentially the same as for on-axis correction. Wider angular aperture results in increased spherical aberration which is more difficult to correct. Even when, as was the case in Abbe and Löber's systems, aberration was corrected for a zone midway between center and edge, there would still be unacceptable residuals when apertures got too wide.

[72] There is a vast secondary literature on Ernst Abbe and the Zeiss company which takes this line – most of which was written by former students of Abbe's and by employees of the Zeiss concern. See F. Auerbach, *Das Zeisswerk und die Carl-Zeiss-Stiftung in Jena: Ihre wissenschaftliche, technische und soziale Entwicklung und Bedeutung* (Jena: G. Fischer, 1903); F. Auerbach (1918); P.G. Esche, *Ernst Abbe* (Leipzig, 1963); W. Mühlfriedel and E. Hellmuth, "Geschichte der optischen Werkstätte Carl Zeiss in Jena von 1875 bis 1891", *Zeitschrift für Unternehmensgeschichte* 38 (1993): 4–25; M. v. Rohr, "Geschichte" (1936–1938); M. v. Rohr, "Arbeitgemeinschaft, I" (1936–1938); M. v. Rohr, "Arbeitgemeinschaft, II" (1936–1938); M. v. Rohr, "Ernst Abbe als Leiter der Werkstätte bis zu seinem Tode, III", *Forschungen zur Geschichte der Optik: Beilage der Zeitschrift für Instrumentenkunde* 2 (1936–1938): 295–350; v. Rohr (1940); F. Schomerus, *Geschichte des Jenaer Zeisswerkes, 1846–1946* (Stuttgart: Piscator, 1952); W. Schumann, et al., *Carl Zeiss Jena Einst und Jetzt* (Berlin: Rütten & Loening, 1962).

[73] L. Dippel, "Mikrographische Mitteilungen", Max Schultze's *Archiv für mikroskopische Anatomie* 9 (1873): 801–12, on 811–12.

[74] See, for example, Carl Zeiss, "No. 23. Illustrierter Katalog über Mikroskope und Nebenapparate aus der optischen Werkstätte von Carl Zeiss in Jena", 1878, BACZ 30528, which cites the correction of aberrations for all zones as the main advantage of his objectives. This language continued to appear after the introduction of homogenous immersion the following year, but with increasing attention given to aperture and resolving power. See also Boegehold (1951), 8–9.

[75] Abbe to Dippel, 9 Oct 1873, BACZ 20386.

[76] "Vier Handschriftlicher Notizbücher über Fertigung von Immersionssysteme", 1873–1893, HStA Weimar, Carl Zeiss Jena, I/1/12323. These notebooks contain what appear to be logs from the mounting and correction of dry F and immersion systems. Because of their relatively short focal length, these objectives could be corrected for spherical aberration only for use with a cover slip of a specific thickness. Thicknesses for which the objectives are recorded as having their best correction range over approximately 25%; homogenous immersion objectives in later entries show a slightly varying numerical aperture. In addition, each objective was evaluated by the tester and given a rating of either "*gut*", or "*sehr gut*." The occasional gap in the record perhaps represents an objective that did not rate the "*gut*." Obviously, some finished objectives were better than others. See also Wittig (1989), 77–78.

[77] Abbe claimed that this variation was only due to frequent changes in the details of the designs. Abbe to Dippel, 21 Jan 1879, BACZ 20386.

[78] The correspondence with Schwendener appears no longer to be extant, but is mentioned in Auerbach (1918), 281.

[79] Abbe to Dippel, 25 Mar 1873, BACZ 20386.

[80] Abbe, "Beiträge" (1873), 93–6; E. Abbe, "Ueber einen neuen Beleuchtungsapparat am Mikroskop", Max Schultze's *Archiv für mikroskopische Anatomie* 9 (1873): 469–80, in *Gesammelte Abhandlungen*, vol. I, 101–12.

[81] Abbe to Dippel, 9 Oct 1873, BACZ 20386.

[82] See, for instance Dippel (1869), 291–2, where he describes a plate of diatoms arranged in increasing order of difficulty, commercially supplied by J.D. Möller for the purpose of testing the resolving power of microscopes.

[83] In one review, Dippel had directly compared it to Zeiss' F, for instance. See quote from Dippel, "Mikrographische Mitteilungen" (1873) above.

[84] See Abbe, "Beiträge" (1873), 91, 99.

[85] Pages 89–91 of Abbe's *Gesammelte Abhandlungen*, vol. I.

[86] Abbe to Dippel, 9 Oct 1873, BACZ 20386. Emphasis Abbe's.

[87] Abbe to Dippel, 23 Oct 1873, BACZ 20386.

[88] Abbe to Dippel, 23 Oct 1873, BACZ 20386.

[89] Abbe to Dippel, 23 Oct 1873, BACZ 20386.

[90] L. Dippel, "Die neuen Objektivsysteme von Carl Zeiss und Professor Abbes Beleuctungsapparat", *Flora* 56 (1873): 497–503.

[91] L. Dippel (1873).

[92] C. Nägeli and S. Schwendener, *Das Mikroskop, Theorie und Anwendung desselben* (Leipzig: Engelmann, 1877), iii.

[93] Nägeli and Schwendener (1877), 84–6.

[94] Nägeli and Schwendener (1877), 87.

[95] Nägeli and Schwendener (1877), 147–9, 156–9.

[96] E. Aberhalden to M. v. Rohr, 22 Mar 1939, BACZ 1012, 112.

[97] v. Rohr, "Zur Geschichte" (1936–1938), 35.

[98] Auerbach (1926), 14; Gesellschaft deutscher Naturforscher und Aerzte, "Sitzungen der XXXII. Abteilung (Instrumentenkunde)", *Verhandlungen der Gesellschaft deutscher Naturforscher und Aerzte* 64:2 (1891): 567.

[99] S. Czapski, "Die voraussichtlichen Grenzen der Leistungsfähigkeit des Mikroskops", *Zeitschrift für wissenschaftliche Mikroskopie* 8 (1891): 145–55.

[100] E. Abbe, "Die Diffraction des unpolarisierten Lichtes und ihre Anwendung auf die Abbildung im Mikroskop", 1893, HStA Weimar II/43/22601 and BACZ 27190. Disappointed that Abbe never published this work, Otto Lummer and Fritz Reiche published a volume based on their notes from Abbe's seminar more than twenty years later. See O. Lummer and F. Reiche, *Die Lehre von der Bildentstehung im Mikroskop von Ernst Abbe* (Braunschweig: F. Vieweg, 1910).

[101] S. Czapski *Theorie der optischen Instrumente nach Abbe* (Breslau: E. Trewendt, 1893); S. Czapski and O. Eppenstein, *Grundzüge der Theorie der optischen Instrumente nach Abbe* (Leipzig: J.A. Barth, 1904); S. Czapski, "Kollegheft: Geometriche Optik n. Prof. E. Abbe, I", 1885, BACZ 11263.

[102] "Protokoll vom 4. August 1903, Sitzung der Geschäftsleitung", BACZ 23013; "Protokoll vom 8. August 1903, Sitzung der Geschäftsleitung", BACZ 23013; "Protokoll vom 11. September 1903, Sitzung der Geschäftsleitung", BACZ 23013; Ambronn to Vollert, 30 Aug 1903, BACZ 23013; E. Crawford, J.L. Heilbron and R. Ullrich, *The Nobel population, 1901–1937: A census of the nominators and nominees for the prises in physics and chemistry* (Berkeley: Office for History of Science and Technology, 1987), 38–9. The first volume of Abbe's *Gesammelte Abhandlungen,* containing his papers on the theory of the microscope (but excluding some of his English contributions,) appeared in 1904. Other volumes followed after Abbe's death.

[103] Rayleigh "On the theory of optical images, with special reference to the microscope", *Philosophical magazine* (1896): 167–95.

DAVID CAHAN

THE ZEISS WERKE AND THE ULTRAMICROSCOPE: THE CREATION OF A SCIENTIFIC INSTRUMENT IN CONTEXT

I presented a first draft of part of this essay at a conference on "Instruments and Institutions: Making History Today" at The Science Museum, London (April, 1991); I thank Robert Bud and Susan Cozzens for their invitation to do so, and John Krige for his helpful comments and criticisms at the conference. I also presented first drafts of other parts of this essay at History of Science Society meetings in Madison (November, 1991) and New Orleans (October, 1994). Finally, I thank Jed Buchwald and, especially, Gerard L'E. Turner for their many instructive comments and criticisms.

I. INTRODUCTION

Virtually all scientists observe the natural world almost exclusively through the use of instruments and all weigh and measure what they see with instruments. The discovery of nearly all new physical and much new biological phenomena; the locating and sizing of new phenomena in the world and in relation to other phenomena; and much of the analysis of measuring results about such new phenomena depends on instrumentation. From the simplest rulers for determining the lengths of objects to the most sophisticated supercomputers for analyzing data, instruments stand at the heart of the modern scientific enterprise.[1]

At the same time, the scientific enterprise has itself become increasingly interwoven with the enterprises of technology and industry. Particularly since the rise of the modern science-based chemical industry in the 1840s and 1850s, and its counterparts for optics and electricity in the 1870s and 1880s, the triumvirate enterprises of science, technology, and industry have become intimately connected with one another. It has thus become increasingly difficult if not impossible for either the historical actors or their historians to locate where one enterprise ends and another begins. This blurring of enterprises is in good measure due to the increasingly dominant role played by instrumentation within and among the enterprises of science, technology, and industry.

Much of the leadership in the science-based chemical, optical, and electrical industries of the nineteenth century first arose and developed in Germany. The reasons for that leadership are essentially twofold. Reacting in good part to the political destruction wrought by the Napoleonic Wars and to the relatively poor supplies of natural resources in Central Europe, the individual German states turned to educating and training their citizenries as a compensatory means of addressing their political insecurities and material shortcomings; education and training lay at the heart of the German response to the dual political and industrial revolutions forged by the French and the British from the late-eighteenth century onward. After the Congress of Vienna, the German states gradually modernized their educational systems: they established mandatory primary schooling; an array of secondary institutions ranging from the classical Gymnasium to the vocationally oriented *Realschule*; a highly complex, varied, and competitive set of institutions devoted to technical education, including intermediate technical schools (*Technische Mittelschulen*) and various sorts of higher trade and polytechnical institutions (*Gewerbeschulen* and *Polytechnische Schulen*) that, after 1877, evolved into the *Technische Hochschulen* (institutes of technology); and, finally, a set of universities that, especially after around 1840, became increasingly devoted to research and equipped with ever more and ever larger scientific institutes. Imparting *Bildung* and *Ausbildung* – cultivation and training – to a significant portion of their populations became the goals of the German *Länder*'s educational systems. Germany's political and industrial leaders believed – and still believe – that an educated and trained cadre of scientific and technical workers could compensate Germany's relative lack of natural resources by using science and technology to process imported raw materials into new and better-finished products. Germany could and did thus use science and technology to boost its economic growth and secure its political sovereignty. At the same time, after 1871 the *Länder* and, to a far lesser extent, the Reich slowly established a number of specialized governmental, non-academic research institutes devoted to such subjects as meteorology, geology, health, agriculture, physical standards and measures, and so on. (A number of private or semi-private research institutes were also established by various industrial associations, such as the sugar-beet industry.) To a large extent, the beliefs of Germany's leaders proved well-founded: from an importer fearful of potential enemies beyond its multisided, unprotected borders, Germany became (after 1880) an exporter feared by its industrial and political rivals and, overall, Europe's leading economic power.[2]

This essay seeks to illustrate how these larger themes of modern German educational and industrial development formed much of the background or hidden foundation – the conditions of producing scientific instrumentation and, with it, knowledge and wealth – that underlay the origins and early use of a specific scientific instrument, the ultramicroscope, one of the most highly sophisticated instruments during the first three decades of the twentieth century and one that played a direct role in the awarding of one Nobel Prize and an indirect role in that for three others. In so doing, the essay seeks to show how the creation of the ultramicroscope – strictly speaking, the first or slit ultramicroscope – occurred in a particular industrial setting and in response to a particular set of scientific problems whose solution had potential implications for achieving fundamental understanding about the nature of matter and fundamental importance for medicine, as well as for potentially increasing profits in a host of industries. The setting was the Zeiss Werke in Jena, the world's foremost optical concern; the problems those of colloid chemistry. The essay argues that full understanding of the origins of the ultramicroscope first requires understanding the broader yet highly specific scientific, technical, industrial, and economic context out of which it emerged.[3] Section 2 of this essay thus briefly recounts the origins of scientific microscopy as it concerns the early years of the Zeiss Werke (1846–66) and some of the salient theoretical results achieved by Ernst Abbe through the early 1870s. Not least among those achievements were Abbe's discovery of the theoretical limit of resolving power of a microscope, a limit that at first blush seemed to preclude the hope of ever imaging or measuring colloidal particles lying below it. The essay then turns (Section 3) to a discussion of the nature and rise of the Zeiss Werke and of the associated glassmaking works of Schott & Co. during the last quarter of the nineteenth century. It shows how the Werke's physical facilities, highly trained scientific personnel, financial wherewithal, and, not least, the attitudes of its owners and management – above all, Abbe – towards research and development provided the necessary material and financial resources and a fitting cultural ethos within its industrial setting that in time led to the creation of the ultramicroscope, the first of which was built at the Zeiss Werke. Sections 2 and 3 (Part One) thus constitute a *mise en scène* in which the broad and hidden context behind the ultramicroscope's creation and development at the Zeiss Werke is presented.

Part Two is devoted to the creation and first uses of the ultramicroscope. Section 4 briefly relates the emergence of the new field of colloid chemistry, some of the scientific-technical problems that it confronted in the 1890s, and the early career of the colloid chemist Richard Zsigmondy. Section 5 goes on to provide a detailed description of the ultramicroscope, conceived in 1901 by Zsigmondy and invented in 1901–2 by him and his collaborator at the Werke, the microscope instrument maker Henry Siedentopf. Here, again, this section emphasizes the industrial context that fostered the creation and initial use of the ultramicroscope as well as the role of advanced instrumentation in scientific change. Section 6, after briefly recounting the well-known innovations in the molecular theory of Brownian motion by Albert Einstein and by Marian von Smoluchowski and the use of the ultramicroscope by two other scientists, The Svedberg and Jean Perrin, to test experimentally Einstein's and von Smoluchowski's theories, offers some concluding remarks concerning the historical development of the ultramicroscope, and of instrumentation in general, in their scientific and industrial contexts.

PART ONE

2. CARL ZEISS, ERNST ABBE, AND THE RISE OF SCIENTIFIC MICROSCOPY

During the eighteenth and much of the nineteenth centuries Britain, which is to say London, and (to a lesser extent) France, which is to say Paris, were the centers of optical instrument making. German scientists who sought high-quality optical instruments normally imported them from abroad.[4] Although Joseph Fraunhofer (1787–1826) in Bavaria did much to raise the quality of German optical instrumentation during the 1810s and the first-half of the 1820s, his death at an early age left the Germans still much dependent on the

British and the French for high-quality optical instrumentation. In 1857, for example, the physiologist Ernst Brücke told Hermann von Helmholtz that Karl Steinheil, one of Germany's leading optical instrument makers, maintained "that the English are far ahead of all Germans".[5] After 1870, however, three individuals championed the transformation of the German optical industry: the instrument maker Carl Zeiss, the physicist and industrialist Ernst Abbe, and the chemist and glass manufacturer Otto Schott. Although each of these three individuals made distinct scientific, technical, and industrial contributions to the transformation of the German optical industry and to the creation of new types of high-quality instrumentation, it was above all Abbe who provided the dynamic intellectual and industrial leadership. However, that leadership, and the subsequent contributions of Schott, can only be understood within the institutional structure first established by Zeiss.

Carl Zeiss (1816–88) founded his mechanical and optical instrument making workshop in Jena in November, 1846.[6] Zeiss himself had little formal scientific training; like most instrument makers of the day, he first learned his craft during a training period with a master mechanic: in his case, between 1834 and 1838 with Friedrich Körner in Jena. However, Zeiss took the unusual step of also attending lectures at the University of Jena. Moreover, between 1838 and 1845 he continued his training by working, according to his own account, in the "most renowned physical, optical, mathematical, and machine workshops in Stuttgart, Darmstadt, and Vienna". "Hereby", he added, "I did not fail to use every opportunity available to the mechanic for acquiring the useful and necessary supporting sciences and arts for my further development". Indeed, in 1845 Zeiss ended his *Wanderjahren*, returned to Jena, and continued to study chemistry and advanced mathematics.[7] Although Zeiss perhaps never warranted the title "scientist", he nonetheless sought to learn as much science as possible so as to enhance his knowledge of instrument making. Moreover, he sought to meet University of Jena scientists, including the botanist Matthias Jacob Schleiden, one of the founders of cell theory, for whom Zeiss built microscopes, both of the simple and the compound variety.

Zeiss began his business with working capital of (a mere) 100 Taler and in rented quarters. At first, fine mechanics rather than optics was his specialty. He made various laboratory articles, scales, eye glasses, and so on, and repaired instruments and equipment belonging to the university's natural science professors. Schleiden, however, urged Zeiss to concentrate on the general area of optical instruments and, in particular, on building microscopes for himself and others. By the mid-1850s Zeiss's workshop, though still quite modest in size – he normally employed only one or two men – began to bring him a modest reputation for his microscopes, and optical instruments replaced mechanical instruments as the heart of his business. By 1857, his business had prospered sufficiently for him to purchase his own quarters.

While Zeiss's business experienced impressive growth from the mid-1850s to the mid-1860s, Zeiss felt the futility of the traditional trial-and-error approach to lens making; he sought a more rational approach. On the one hand, thanks to a new compound microscope that he built during this period, by 1866 he could count 20 employees who helped construct 146 compound microscopes that year. His own reputation grew, too: in 1860, he was appointed university mechanic; and in 1861, competing against 1,300 others, he won a gold medal for his microscopes. On the other hand, Zeiss's passion to create the finest microscopes, and to do so on the basis of scientific theory and according to pre-calculated mathematical specifications for cutting and shaping lenses and other optical components, remained unfulfilled. He turned to textbooks in optics for guidance and was prepared to abandon artisanal tradition in microscope construction if he could discover a better approach. Yet he found that he lacked sufficient scientific training to master and develop practically what he read. Moreover, he believed that if he did not achieve further technical innovation for his microscopes (especially for the objectives) his competitors would eventually drive him out of business.

Hence, in 1866 Zeiss sought to ensure his business's future by hiring a scientific consultant, the young University of Jena physicist Ernst Abbe (1840–1905). From 1846 to 1866 Zeiss had been not only the founder and sole proprietor of his firm, but also its master. Practitioner now let himself be guided by theorist – though one, to be sure, whose ultimate aim was to design far better and more economical instruments. The result was not only the transformation of Zeiss's small firm, but also an essential moment in the rise of scientific microscopy and of Ernst Abbe.

Zeiss first made Abbe's acquaintance when the latter was a physical science student at the University of Jena, and he immediately came to appreciate Abbe's knowledge of theoretical physics. No doubt Zeiss saw in Abbe someone who possessed precisely the sort of rigorous, formal training in physical science and mathematics that he himself lacked. When Abbe's university education began in 1857, Jena was a small, economically undeveloped university city in the poor state of Thuringia; the university, accordingly, was impoverished.[8] For example, the university then had only *one* ordinary professor for the combined fields of mathematics, physics, and astronomy. It had neither an institute for physics nor permanently assigned lecture rooms for physics; its instruments and equipment were primitive and outdated. Moreover, Abbe's teachers, the ordinary professor Karl Snell and the lecturer Hermann Schaef-fer, offered neither experimental physics nor laboratory work.[9] Apart from a course by Snell on analytical optics, Abbe's only contact with practical optics came through his acquaintance with Zeiss and his small instrument making firm.[10]

Dissatisfied with Jena's poor, inadequate scientific offerings, Abbe trans-ferred in 1859 to the University of Göttingen. There he found some of the era's finest physicists, mathematicians, and physics facilities. In particular, he found his principal, most influential teachers: the physicist Wilhelm Weber and the mathematician Bernhard Riemann. Two years after entering Göttingen, he graduated with a dissertation on the experimental foundations of the theory of the equivalence between heat and mechanical work – and with Weber's high praise.[11]

In 1863, Abbe habilitated at Jena (with a study of error distribution in experimental data and the method of least squares) and so became a *Privatdozent* there. His lectures, at least in his first years, were usually attended by only one or two students, and by never more than a dozen; he was a poor lecturer. Though he initially taught several different types of courses in both mathematics and physics, he later came to limit his lecturing to particular aspects of optics. His laboratory course in experimental physics, by contrast, proved more successful; during various semesters he attracted around seventeen students. During his first years of teaching, Abbe had only the most primitive, outdated instruments at his and his students' disposal; and his so-called laboratory was, in fact, a mere shed, which served until 1880 when it collapsed during the course of a visit by the zoologist Ernst Haeckel![12] The material problems of teaching experimental physics at Jena only added to Abbe's burden of a heavy teaching load. In a letter of 1863 to his close friend Harald Schütz, he discussed his course on the theory of measuring instruments and in so doing revealed the handicaps of teaching physics at Jena:

But then came the experiments. At first I scarcely reckoned with them because the physical cabinet has almost no noteworthy measuring instruments with which I could have demonstrated. I soon saw, however, that a theory of measuring instruments, one so completely abstract and without accompanying demonstrations of the apparatus described and without practical exercises, is something damned boring, and that I would earn little approval for my lecture course if I could not join practical demonstrations to it. ... Day after day I have thus had to worry about the work [of constructing or having constructed new demonstration apparatus]. Moreover, this needs to be done as quickly as possible if it is to fit into my current lecture course. ... You've never been concerned with physical experiments and so you have no idea of the time and trouble it costs when one is supposed to put together a tolerable piece of apparatus with such inadequate resources and with one's own hands.[13]

Both as a student and as a young lecturer, the shortcomings of the University of Jena had made clear to Abbe the need for good instrumentation and good laboratories for both teaching and research purposes. It was this need for good instrumentation that above all drew him to Zeiss.[14]

During his first seven years at Jena (1863–70), Abbe was so preoccupied with teaching and with practical work in optics that his scientific research never reached a finished state; before 1870 he did not publish a single line beyond his *Habilitationsschrift* of 1863. At his first attempt, in 1869, to be promoted from *Privatdozent* to extraordinary professor, he was therefore rejected. While he was promoted in the following year, it would take him nearly another decade until, following an offer from Berlin (extended personally by Hermann Helmholtz in May 1878 during a visit to Abbe in Jena) to come to the university there as a special professor of optics, Abbe was again promoted (July 1878), this time to honorary ordinary professor.[15]

Although Abbe did not publish before 1870, he was nonetheless intellectually active and creative. At the heart of his scientific efforts stood the goal of establishing scientific foundations for microscopy, aimed at improving the quality of the microscope as well as its methods of production. Abbe and his employer Zeiss resolved to abandon the traditional trial-and-error methods of lens construction: Abbe sought to replace those methods with scientifically based predictions about the behavior of light rays within lens systems. He and Zeiss held to their program so strongly that (at first) they cared less about improving lenses *per se* than about ensuring that their craftsmen properly executed the instructions given to them on the basis of Abbe's theory and calculations. Hence, they innovated not only by seeking to replace trial-and-error methods with mathematically and scientifically based methods (which themselves were hardly error free), but also by introducing a *division of labor* in microscope production. Microscopes at Zeiss were no longer, as in the past, built by one single instrument maker; instead, each craftsman was responsible for constructing only a specific part of a microscope according to the instructions given him by Abbe. It was this new organization of work, as much as its mathematico-scientific basis, that was so distinctive about the Zeiss Werke. Abbe coordinated all these individual efforts into producing a finished product that fully met his optical specifications.[16]

In attempting to improve microscope lenses, Abbe began by assuming that traditional geometrical optics, in particular the views of David Brewster, provided a solid basis for optical theory. Before Abbe, it was believed that the image-formation process in a microscope virtually reproduced the object's structure and that it could do so to an unlimited degree of magnification. The pre-Abbean theory of the microscope, such as it was, was largely based on the telescope. It took into account only the geometrical laws of light reflection and refraction; only occasionally did physical optics enter into consideration: diffraction of light was only a minor consideration. Yet telescopes are, of course, designed to magnify distant, self-illuminating objects. By contrast, the microscope seeks, of course, to view small, close by, non-self-illuminating objects. To see microscopic objects thus requires illuminating them and contrasting their brightness with their surrounding environment.[17]

Pre-Abbean theory also held that to form the best optical image required an objective with a small angle of aperture. When Zeiss craftsmen, following Abbe's strict instructions, produced small-angle objectives, they found that the images formed by these objectives showed less brilliance and lower resolving power than those formed with large-angle apertures. Abbe's explanation – and the essence of his revolutionary understanding of the behavior of light rays in microscopes – concentrated on the influence of the objective's aperture on resolving power. He argued that diffracted light rays are produced by closely spaced structures on the object being observed (these structures acting, in effect, as a diffraction grating); these rays, Abbe further argued, are dispersed around the sides of the axial illuminating beam. He thus recognized that to develop a theory of the microscope meant understanding the behavior of diffracted as well as of reflected and refracted light. The amount of this dispersion, he found, is a function of the periodic structure of the object; in particular, the more closely spaced the object's structures, the greater the degree of separation of dispersed light rays. Abbe thus reasoned that to increase resolution required objectives with apertures sufficiently large so as to include at least some of the diffracted light rays (as well as those of the axial beam). To capture the diffracted rays, he called for the use of oblique illumination and large-angle objectives. Furthermore, he found that in order to obtain a uniform image the different points on an object's surface had to be magnified equally. This in turn required that two conjugate points – i.e., a pair of points such that all rays from one will image onto the other – have a constant ratio between the sines of their respective angles at the aperture. The fulfillment of this condition, subsequently known as Abbe's "sine condition", allowed the imaging of both axial and diffracted light rays. Abbe's theory of image formation thus concerned the diffracted as well as the geometrical image.

No sooner had Abbe created his theory then he sought to reduce it to a formula for microscope design and for determining the theoretical limits of resolving power. He called his central concept for measuring the limit of resolving power the "numerical aperture", that is, the capacity of a lens to accept light. Moreover, he found that three factors limited resolving power: the wavelength λ of the illuminating light; half the objective's angular aperture α; and the refractive index n of the medium between the object and the objective. The smallest distance d capable of being resolved between structures was thus

$$d = \frac{\lambda}{2n\sin\alpha}.$$

Abbe denoted the term $n\sin\alpha$ as the numerical aperture NA, or $NA = n\sin\alpha$. Hence

$$d = \frac{\lambda}{2NA}.$$

Hence, further, the limitation on distinguishing microscopic structures was a function of the wavelength and the numerical aperture, and with sufficiently small wavelength and sufficiently large aperture one could, in principle at least, increase resolving power as far as practicable. Thus, a lens with a numerical aperture of, say, 1, would have twice the resolving power of a lens with an aperture of 0.5. Abbe settled on the half-wavelength of the illuminating beam as a practical limit to magnification, since numerical apertures could not easily exceed unity. Light could not, on this basis, resolve objects below about 0.25 μm, which in turn implied, to anticipate, that one could not possibly image molecular let alone atomic phenomena. By 1872, after some six years of labor, Abbe had developed his theoretical system of image formation and its accompanying formulas. Though he would continue to add theoretical and empirical refinements thereafter, by then his theory of the microscope was essentially complete.[18]

3. THE ZEISS WERKE AND SCHOTT & CO. (1872–1905)

In effect, then, Abbe's arrival at the Zeiss Werke in 1866 marked the founding of an industrial research laboratory there. For Abbe and his employer Zeiss were at least as interested in the practical realization of significantly improved microscopes and microscope apparatus as they were in Abbe's achievement of theoretical understanding of microscope imaging. Already in 1869 Abbe had developed his substage condensing lens – known in Britain as "the Abbe illuminator"; by 1872, the Zeiss Werke began marketing it. In 1871, optical craftsmen at Zeiss began producing microscope lens systems that met Abbe's precise theoretical conditions. In 1874, the Werke also began marketing Abbe's refractometer for determining the refractive index of either solids or liquids and, at about the same time, Abbe's apertometer for measuring the objective's numerical aperture. In 1872, in an effort to provide a uniform refractive index from the objective to the cover glass above the object, the Werke began building microscopes with water-immersion systems; by 1879, following a suggestion given by the English microscopist J. W. Stephenson, it began regular marketing of its homogeneous oil-immersion lens system. That system permitted all light rays to pass through a uniform medium possessing only one refractive index; it thereby eliminated the air between the covering glass and the objective, and hence produced an environment with one refractive index, that of the covering glass and objective. Moreover, the Zeiss Werke created, again thanks to Abbe, a host of other optical measuring instruments – for example, the focimeter (focal length), the spectrometer (refractive index and dispersion), the sphero-meter (thickness), and comparator (for comparing the precision of two microscopes) – that allowed the rapid measurement of individual lens proper-ties and lens systems. Thus, by the 1870s the Werke was producing practical, everyday microscopes of the highest scientific and technical quality based upon Abbe's well-tested scientific theory. (From 1880 or so onward, Zeiss micro-scopes became increasingly favored by Continental, especially research, micro-scopists; indeed, even some of their British counterparts began using them, too.) The entire process of microscope (and other optical instrumentation and apparatus) production became rationalized: lenses and other optical parts were cut to meet Abbe's theory; testing instruments and apparatus were developed to ensure that a constructed part met ever-more-precise optical and mechanical standards; and the division of labor was increased.[19] The Zeiss Werke never abandoned this controlled, precision scientific mode of operation.

These theoretical and technical innovations brought unprecedented and impressive growth and financial success to the Zeiss Werke. While many German and other European firms experienced the hard economic times brought on by the Great Depression (1873–96), the Zeiss Werke flourished as never before. The period from 1866 through 1885 constitutes the second developmental phase in the firm's history, for by the latter date Abbe had completed his image-formation theory (1872–73) and its results began to manifest themselves in microscope production in the following years. Between 1856 and 1870, for example, the firm produced a total of 1,237 microscopes.

Then, following modest growth in the early 1870s, in the fiscal year 1875–76 alone the firm produced an unprecedented 556 microscopes. Though there was a decline in production the following year, by 1879–80 production had gradually risen back to 558 units; and by 1885–86, it reached an unprecedented high of 1,459 units. All the while Zeiss and Abbe reinvested their profits in the firm (see below).[20]

The start of Abbe's consultancy in 1866 thus led to the gradual transformation of the workshop of Carl Zeiss into the factory known as the Zeiss Werke. Zeiss naturally recognized Abbe's past accomplishments and future importance for the Werke. In 1876, with the firm valued at some 66,000 marks, Zeiss made Abbe a silent partner with an equity of one-third and he transformed the workshop into an incorporated business entity. Although Abbe continued to teach at the university until 1893, his academic career now took a decidedly second place to his industrial career.[21] By 1880, he had become the dynamic leader of a medium-size firm employing about eighty workers.

Despite the success of Abbe personally and of the Werke as a corporation, one key component needed to achieve increased resolving power and unsurpassed microscope quality continued to elude and, hence, frustrate Abbe and Zeiss. From the start of their association, the two recognized that no matter how advanced Abbe's scientific theory became and no matter how well Zeiss employees managed to embody his theory and calculations in their lenses and other instrument parts, there remained one dimension of microscope production that at once lay at the heart of their enterprise and yet completely beyond their control: the composition and quality of the glass used to make lenses. Abbe wrote:

For years we pursued . . . a so-called phantasy optics along with a real optics, [wherein we imagined] designs with hypothetical glass that did not at all exist. In the meantime, we discussed the progress that would be possible if the producers of raw material could ever become interested in the advanced problems of optics.[22]

The reason for their frustration here was simple: glass manufacturers had little or no economic incentive to improve the quality of glass used in optical instruments. In comparison to the large market for ordinary window glass, glassware, and so on, that for scientific instrumentation was minuscule. The glass industry found the market for scientific instrumentation (essentially microscopes and telescopes) simply too small to merit research in and development of improved objectives and lenses.

In particular, Abbe and the Zeiss Werke needed improved, high quality, and highly specialized glass in order to eliminate or at least minimize the two sorts of well-known aberrations that led to image defects and that had traditionally plagued optical instrument makers: chromatic aberration, wherein color fringes, as a consequence of the varying extent to which a glass lens refracts the various wavelengths of light, surround the image; and spherical aberration, wherein rays originating at different distances from the main axial beam focus, after undergoing refraction through a spherical lens surface, at different distances along that beam. Of these two sorts, it was the former that caused Abbe and the Werke the most trouble – and whose correction provided the greatest rewards.

During the first years of his consultancy, Abbe had introduced the use of achromatic lenses to correct some chromatic aberration. His success here was only partial; although achromatics – that is, the combination of two different lenses composed of two different kinds of glass (e.g., a positive lens of crown glass with a negative lens of flint glass, with each lens being of different refractions and dispersions) – noticeably reduced the light's dispersion, they did not eliminate its chromatic abberration: some residual color – known as the "secondary spectrum" – remained. It was not possible to eliminate the "secondary spectrum" without reducing the numerical aperture to an unacceptable extent. The fundamental difficulty lay, Abbe soon realized, in the chemical composition of the two kinds of glasses then exclusively in use: crown glass, which gave weak refraction and weak dispersion, and flint glass, which gave strong refraction and strong dispersion. Abbe's goal, therefore, was to find a means of correcting both spherical and chromatic aberration while still maintaining a large numerical aperture. In 1874, if not earlier, he had begun searching after or calling for experimental work in glass technology (melting) aimed at producing a new type of optical glass designed to meet the needs of his optical theory.[23] In particular, he needed, as he said in 1876, optical materials that combined either "low refractive index with high dispersion or high refraction with relatively low dispersion". In this way he anticipated eliminating much of the chromatic and spherical abberations independently of one another.[24] The problem thus became one of finding a glassmaker who had a good scientific understanding of glass chemistry, who shared Abbe's scientific and industrial–technological goals, and who would follow his scientific specifications. In 1879, Abbe found his man in Otto Schott.

More precisely, Schott found Abbe. As the latter soon learned, Schott (1851–1935) was the son of a glassware maker.[25] He had studied chemistry and chemical technology, first at the Aachen Polytechnic, then at the universities of Würzburg and Leipzig, and finally at Jena, where in 1875 he received his doctorate with a dissertation entitled "Die Fehler bei der Fabrikation des Fensterglases" ("Errors in the Production of Crown Glass"). His family background in the glass business, his advanced scientific and technical training, and his travels to foreign glassworks in France, Belgium, and Spain filled Schott with a burning desire both to extend knowledge of glass chemistry and to use that knowledge to make major innovations in glass production.[26] Schott was thus to glass chemistry what Abbe was to microscopy: a university-trained scientist bent first on comprehending on the theoretical level the material objects with which he worked and then on producing improved versions of those objects on an industrial scale.

Schott initiated contact with Abbe in May of 1879: he reported that he had created a new, improved type of lithium-based glass that showed an excellent ratio of refraction to dispersion.[27] Though impressed by Schott's initiative and results, Abbe knew that Schott had yet to create the right type of glass for improved optical instrumentation. During the next eighteen months, Abbe and Schott corresponded about the need and optimal conditions for producing improved optical glass on the basis of methodical chemical analysis.[28] Late in 1880, Abbe wrote to Schott about some of the more practical concerns that he also had. His letter merits extensive quotation, for among other things it reveals some of the economic hindrances behind scientific and technological change and the industrial and political context in which such change might occur. Of Schott's proposed research and development program to produce a new type of glass, Abbe wrote:

I would do everything in my power to support you in the pursuit of the same. However, I cannot conceal that in my estimation the realization [of such a program] cannot but have a few difficulties – namely, it will demand a tremendous consumption of capital and, at first, unpaid labor. – That such an undertaking promises no easy success can already be seen by the fact that, despite the steep increase in the consumption of optical glass during the past decades, production has remained completely in the hands of a few old factories in Paris and Birmingham. In any case, so much follows therefrom that such undertakings – considered from a business point of view – must have serious difficulties. As for the incentive to get involved in this field, there certainly is no lack [of it]: the cheapest optical glass, crown and flint, as used for example in photographers' objectives, costs at [the Paris glass factory of] Feil 10 to 12 francs per kilo, and in Birmingham [it costs] more than double that, with the heavier flint glasses climbing up to a price of 60 francs and more. – Above all, it will demand much in the school of practical experience just to attain what Feil and [the Birmingham glass factory] Chance [Brothers & Co., Britain's sole glass manufacturer] already presently offer in the way of available types of glass, and thus to become competitive with them in quality and price.

As concerns the consumption or demand for optical glass, from the commercial point of view there is, at present, consumption only by optical factories which produce eyeglasses, hand telescopes, opera glasses, and some photographic apparatus. In this branch alone is there mass consumption. ... Glass for astronomical purposes is indeed purchased at frightfully [high] prices ...; but the demand for such, especially in Germany, is rather limited.... The manufacture of microscopes naturally requires only tiny quantities, from which a commercial enterprise would thus hardly be able to exist, even if it would succeed in producing special glasses for such purposes for which opticians could (and would) pay arbitrarily high prices .... To obtain foreign capital ... for such an enterprise will present great difficulties, especially in Germany. I have no idea how one can do that. However, a state subsidy should not be entirely excluded. Nonetheless, I believe that it would be completely hopeless to attempt any steps in this direction before there is some indication of a certain guarantee of success. In my opinion, that would be possible if you could succeed either in producing crown and flint glass of the known kind in a competitive manner or if you could succeed in presenting some new sort of glass – even if only in small samples – whose use would offer an essential advantage for optics and make [it seem] probable that production of the same in larger quantities (and at good quality) will succeed. In the former case: to establish the domestic production of optical glass in Germany and to emancipate the German [optical] industry in this matter from foreign [producers] would speak in favor of a public subsidy. In the latter case: a direct advance of practical optics, and thereby an indirect advance of scientific problems, would already justify the [financial] support of a private enterprise.[29]

As this long extract reveals, Abbe, who by 1879 had spent thirteen years in association with Carl Zeiss's firm, had become enough of a businessman to understand that the research and development of new glass could only be pursued with new capital resources and that potential investors would require indications of a reasonable possibility of success (i.e., some or increased market share). He had to break the oligopoly held by Feil and Chance, and to do so he had to make significant financial investments in order to produce superior optical glass. A week later, Abbe wrote a follow-up letter to Schott in which he stressed that he did not mean to discourage Schott from his undertaking – quite the opposite – but, rather, that they needed to convince outsiders of the worth of their cause. The whole problematic of composing and producing new sorts of glass was, he said, a *tabula rasa*: neither the chemical properties of such (hypothetical) glass nor the systems, methods, and facts of such glass production were known. Abbe advised Schott that he would "succeed only through the application of methodical experiments" and by producing "the simplest compounds possible". And above all, he advised Schott to analyze each of his glass syntheses and to determine their optical constants "by exact measurement".[30] Indeed, although practical optics (i.e., the commercial sales of optical instrumentation) was their long-term goal, early in their relationship Abbe advised Schott to think of the work as a pure scientific endeavour. He wrote:

All in all: become used to the thought that the work that lies before you will require a large and long-term investigation in order really to bring it to a proper conclusion; but, if this happens, your experiments will be a milestone in the development of optical glass-making, and therefore in practical optics itself![31]

Schott followed Abbe's advice. Beginning in 1881, in his small laboratory in the town of Witten (near Essen, and far from Jena), he created precisely the sorts of samples that Abbe had called for. In particular, that year he discovered that boric acid and phosphoric acid were the basic elements needed to create a new and better sort of glass for practical optics, and that only the boric-acid glass sufficiently gave the right combination for optimal refractivity and dispersion, that is, the combination that Abbe was calling for. Yet Schott had created no more than a sample (circa 20–30 grams); he had still to prove whether or not such new glass could be produced in a factory and on a sufficiently large scale so as to merit industrial manufacture.[32]

In early 1882, Schott moved to Jena and established a small "Glastechnisches Laboratorium". His and Abbe's overall goal now was to create improved optical glass on a sufficiently large scale so as to be of practical, commercial use. Zeiss, and the latter's son, Roderich, now invested 40,000 marks to help found the laboratory. As in all industrial ventures, major problems soon appeared. For one, Schott's boric-acid glass proved industrially unfeasible. In response, he then devised so-called boric-silicate glasses, which combined boric acid, silicic acid, and various alkalis. In the meantime, he and his associates remained uncertain as to whether or not there existed a sufficient market for his new optical glass. For another, Abbe, Zeiss, and Schott no longer had sufficient venture capital to finance further research, and they saw no prospects of obtaining outside investors.[33] So they did what many such struggling venturers do: they turned to the government for financial support.

Although the Imperial German government sought during the 1880s to pursue a laissez-faire economic policy, it had nonetheless become involved in an effort, led by the Berlin industrial scientist Werner Siemens and the polymath scientist Hermann von Helmholtz, with the support of, among others, the Berlin astronomer Wilhelm Förster, to establish what by its opening in 1887 would be known as the Physikalisch-Technische Reichsanstalt.[34] Abbe was aware of these efforts. In May 1883, he wrote Helmholtz:

In November of last year I had the honor of communicating to you concerning experiments which over the past two years or so I have undertaken, in conjunction with a chemist friend, Dr. Schott, aimed at the rational production of glass for optical and other scientific purposes. Already at that time I permitted myself to ask your kind support in favor of an eventual state subsidy to Dr. Schott for the practical realization of our work. I assume that the deliberations which have taken place since then concerning the steps toward the advancement of scientific technology in general [i.e., the eventual Reichsanstalt] have perhaps offered or soon could offer you an occasion for approaching the said goal. I therefore now permit myself the respectful request that, following the practical realization of results, you grant me an opportunity to report to you somewhat more thoroughly concerning the present state of our work and our plans and, at the same time, to hear your kind advice concerning the latter. Our work has now reached a point where a decision concerning the manner of its winding down cannot be delayed much longer. On the one hand, we believe that we have now made advances on the relevant scientific and technical problems to the point that, in general, it is no longer possible within the scope of laboratory experiments; hence, further continuation of the work on the current scale would no longer be purposeful. On the other hand, the very considerable expenses which the maintenance of a special laboratory and several assistants for Dr. Schott imposes upon me has exhausted my means to the point where I must certainly think about soon closing the existing experimental station. However, I do not want to take this decisive step for the future without first having asked you for your authoritative view as to whether there may exist any prospect for realizing our hope concerning the practical utilization of our work.

Abbe added that he would like to come see Helmholtz in Berlin the following week, so as to discuss this matter in person.[35]

Helmholtz helped Abbe, doing so initially by putting him in touch with Förster, who, as head of the Normal-Eichungs-Kommission, which was charged with establishing and maintaining standard weights and measures, was interested in the advancement of improved thermometer glass. Förster sent his subordinate, Hermann Wiebe, who knew Schott from their schooldays together, to Jena in order to assess the possibility of developing improved thermometer glass. Wiebe soon saw, and in turn convinced Förster of, the practicality of developing a new type of glass. In March 1882, Förster had Abbe and Schott compose a memorandum on the progress of their efforts to date and on their future plans and needs. By July, that memorandum had landed on the desk of the Prussian Kultusminister, Gustav von Gossler.[36] Together, Förster, Helmholtz, Abbe and Schott were establishing a Berlin–Jena thermometric axis.

The Berlin–Jena consortium for developing thermometer glass originally hoped to channel support for Schott's glass studies through the institutional framework of the Reichsanstalt. But delays in establishing the latter and the desire to expedite the development of Schott's new glass led them to seek government support now. The Berlin–Jena effort became an instance of pre-Reichsanstalt support for scientific–technical research not associated with the universities or *Technische Hochschulen*. Gossler and Wilhelm Wehrenpfennig, a leading figure in the Kultusministerium, assented readily to public support for the Jena glass laboratory. Helmholtz was more cautious, but nonetheless equally supportive. In late May 1883, Abbe visited Helmholtz in Berlin and convinced him of the necessity of supporting the laboratory. Abbe wrote to a friend about his visit: "I succeeded thereby in interesting Helmholtz to become active for the cause and to maintain his support, which in Berlin is very considerable".[37] Later Rudolf Virchow, one of Europe's leading medical microscopists and an important liberal politician in the Prussian Landtag, became another ardent supporter of the Jena project. In March 1884, he wrote the Landtag's budget commission of the multiple reasons why it should devote public funds to such a project:

We are here indeed concerned with a national undertaking: the problem of producing, in Germany and in an independent manner, glass that is necessary for all scientific purposes as well as for the [general] population with respect to the production of eyeglasses, opera glasses, and the like. Nevertheless, the latter is not the main goal. Above all, it is much more a matter of the production of glass for telescopes, microscopes, and similar scientific instruments. This undertaking is [also] of very special importance for the production of instruments that serve military and naval purposes, where to date we are completely dependent on foreign countries and where ... only chance has made it possible that in the French Wars [i.e., the Franco-Prussian War of 1870–1] the necessary amount of glass could be procured so as to produce the necessary optical instruments for the army. – The men who are presently involved with pursuing this undertaking have the extraordinary merit of having produced, on the basis of a pure scientific foundation and on the basis of exact mathematical calculations, completely new types of glass from substances other than those which have been used to date, and [they have done so] in a way which promises not merely to cover what we have until now bought from abroad but also to push further ahead in the use of natural substances for the solutions of the most difficult optical problems. – I can only say that scientific interests look forward to this undertaking with the greatest excitement and that the sums which are to be expended here will doubtless be useful to the greatest degree for all industries based upon [the use of] glass.

The budget commission and the Landtag presumably found Virchow's public argument, which is strikingly similar to parts of the private argument that Abbe had advanced to Schott in 1880, convincing: they approved 60,000 marks for two years' worth of further research and development of the new glass. By the fall of 1884, Schott and his partners (Abbe, Carl Zeiss, and Roderich Zeiss) had established a new company – the Glastechnisches Laboratorium Schott und Genossen – and built a new factory just outside of Jena.[38] (See Figure 1.) Thus, five years after he had first begun glass–chemical research aimed at meeting optical properties for glass as specified by Abbe, Schott was ready to enter into commercial production and to reap the financial benefits of a revolution in glass technology that he had conceived and led.

Almost immediately the Schott glass chemical plant began producing a wide variety of glasses meeting the precise needs of Zeiss (and other) optical manufacturers. In 1885, the Zeiss Werke, employing new types of glass produced by Schott, began marketing Abbe's new apochromatic (i.e., "away from color") objective systems, which, along with Abbe's new compensating eyepieces and projection eyepieces, eliminated virtually all chromatic and spherical abberation.[39] These apochromatic systems constituted Abbe's principal achievement in instrumentational microscopy. Virchow's fellow microbiologists – including bacteriologists, microanatomists, histologists, and cytologists – along with any number of others found Zeiss microscopes with apochromatic systems to be particularly useful, for their unprecedented image

Figure 1. The Jena Glassworks in 1886. Source: Rohr, "Zur Geschichte", in *Forschungen zur Optik 1*: 159 (see note 7).

qualities reached the limit of resolution.[40] In July 1886, Schott and Abbe issued a product list of no fewer than forty-four different types of glass for optical and other special purposes, and for each glass type they detailed the exact chemical composition, refractive index, ratio of refraction to dispersion, specific weight, hardness, and other optical properties.[41] (Subsequent product lists would present still more different types of glass for sale.)[42] Then in 1889 Abbe oversaw the completion of his master microscope system: an apochromatic objective with an unprecedented numerical aperture of about 1.63 set within a naphthaline monobromide immersion system. With it, Abbe had reached the theoretical limits of resolving power, and thus of the light microscope itself. At the same time, throughout the 1890s Abbe and the Werke continued to make improvements on their apochromatic microscopes, and so to set the pace in advanced microscope development and associated microscopical researches.[43]

The Prussian government had made a sound investment, and one that quickly paid dividends. With the wide variety of new types of glass now available from Schott & Co., Abbe and his Zeiss Werke now commanded and controlled every aspect of optical instrument making. After 1886, the Werke became the unchallenged master of microscope production – both in terms of quantity and quality – in Germany if not abroad. The Germans, in the institutional form of the Zeiss Werke and Schott & Co., now began producing noticeably better microscopes than the British and French. Where once the Germans (and others) had looked to Britain and France for fine optical instrumentation, they (and others) now looked to the Zeiss Werke in Jena.

Moreover, the Zeiss Werke produced both high-quality optical products for everyday use as well as specialized instrumentation for advanced scientific research, including such items as eyeglasses and opera glasses; lenses for amateur photographers' cameras; photographic objectives; butter and milk refractometers; field glasses, stereoscopic range finders, and sighting telescopes, all of which were important to the military; microphotographic apparatus; photometers; dilatometers; and telescopes. By the late 1880s, Zeiss optical products were the high technology of the day, and the name "Zeiss" on a product was the sign of the finest quality.

The revolutionary development in glass technology and the new or improved products from Zeiss associated with that development, had important consequences for the Zeiss Werke. The latter's history during the first sixty years of its existence (that is, under the leadership of Zeiss or Abbe) breaks down into three twenty-year periods. The first (1846–66) set the basis for the firm's existence, though that basis, to be sure, remained quite shaky and the firm remained quite small. The second (1866–85) was characterized above all by Abbe and Schott's various dramatic technological developments that secured the firm's financial future and turned the workshop into a factory. The third (1886–1905) saw extraordinary financial growth, corresponding expansion of production facilities, and increases in market share that transformed the Zeiss Werke into a large industrial enterprise that, though still located in the small town of Jena, conducted business on a worldwide basis.[44]

Several business indicators illustrate the firm's stunning development after 1885. As financially successful as the period from 1866 to 1885 had been for the Werke, it was not nearly as spectacular as that which followed from 1886 to 1905, the year in which Abbe died. With the advent in 1886 of the apochromatic microscope and, more generally, Schott glass tailored to Zeiss optical instrumentation, the Werke entered a new, unprecedented era of growth. In 1877, the Werke's annual sales were about 120,000 marks; in 1885, about 400,000; in 1886, about 500,000; in 1896, about 1,875,000; in 1900, about 3.15 million marks; and in 1902, about 3.6 million marks. In 1877, the Werke alone (i.e., without Schott & Co.) employed perhaps 40 individuals; in 1885, around 250; in 1896, some 700 individuals; in 1900, over 1,000; and in 1902, over 1,300. In 1866, when Abbe first entered the firm, the Werke sold just over 200 microscope stands; in 1879, about 470; in 1886, slightly more than 1,400; and in 1902, some 1,930. To make a similar point in terms of marks: in 1890/91, the Werke sold no fewer than 992,900 marks' worth of microscopes (the sales were especially high that year due to the aftereffects of Koch's discovery, using a Zeiss apochromatic microscope, of the tuberculosis bacillus). In 1899/1900 it sold 617,200 marks' worth of photographic objectives; by 1906/07, it sold over 1,000,000 marks' worth. Field glasses, which Zeiss first sold in 1894/95, brought sales of 187,000 marks that year; by 1900, sales topped 1,000,000 marks. Non-Germans were so impressed with Zeiss goods that much of the Werke's sales took the form of exports: in 1899, for example, two-thirds of the instruments produced by Zeiss were sold abroad.[45] Where Britain and,

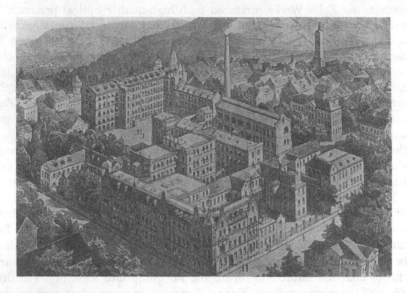

Figure 2. The Zeiss Works in 1905. Source: Rohr, "Zur Geschichte", in *Forschungen zur Optik 1*:198 (see note 7).

especially, France had once dominated the precision instrumentation industry, now they not only gradually lost ground to the Germans, but, thanks above all to the spectacular rise of the Zeiss Werke, after 1890 German precision instrument makers decisively took the lead. Zeiss dominated the international market in optical instrument making.[46]

Befitting as well as necessary for this extraordinary growth in production and sales, the Werke acquired new real estate on the outskirts of Jena in 1879, and during the 1880s it constructed five new buildings (nearly 3,000 square meters) and increased its equipment and facilities accordingly. (By 1905, the Werke comprised about one-quarter of the city of Jena.) During the 1890s additional plant was built nearly every year. By 1905, the Werke occupied more than three acres of real estate, about half of which included factory buildings (covering almost four acres of floor space) in which between 1,300 and 1,400 people worked.[47] (See Figure 2.) And within the new facilities precision-automated machinery increasingly replaced skilled workmen and more scientists or scientifically-oriented engineers (see below) ran the Werke.[48]

From the early 1870s on, the Zeiss Werke's growth and the aging of its patriarch meant that changes in management were needed, too. In 1876, Carl Zeiss's eldest son, Roderich, joined the Werke. Five years later he became a full partner and assumed much of his father's managerial responsibilities. Together with Abbe, Roderich Zeiss helped oversee the Werke's transformation from the old trial-and-error workshop into the new scientifically-based industrial firm. However, following Carl Zeiss's death in 1888, Abbe and Roderich Zeiss, who now became equal equity partners, fell out over the future direction of the Werke; a year later, Abbe essentially forced Roderich Zeiss to retire from the firm's management.[49]

At least as early as the mid-1880s, Abbe wanted to ensure perpetuity for the Werke. As the son of a poor, working-class father, he also wanted to enhance his workers' welfare – to help the men who had helped him and Zeiss build the Werke. More generally still, he wanted to use part of the firm's profits to enhance the public welfare. And finally, he wanted to help the general advancement of science (above all at the University of Jena), an advancement which had done so much for the Werke.[50] Thus Abbe sought not only to secure the Werke's future but also to repay and to help his fellow workers, academic science, and the public weal, each of which had helped the Werke. In his seventy-two-page, handwritten memorandum of 4 December 1887, he explicitly declared his reasons for his plans to establish a public trust:

Among the sundry ways that a man of my way of thinking is inclined to feel himself called to serve the public weal, only two can for me – considering the given personal circumstances and circumstances of time – come into consideration: abundant provision for the well-being of all those who have contributed to the attainment of those means to date or will contribute in the future; and the advancement of the sciences, in whose soil the enterprises involved here have grown and to which, at the same time, I owe thanks for my own rise in the world.

With respect to the first, no further resolution is needed – except for the special measures whose scope is given of itself through the administration of the institutes in question [i.e., the Zeiss Werke and Schott & Co.].

But also with respect to the second, the choice is, assuming I do not want to leave any room for arbitrary inspirations, completely prescribed. Only the University of Jena has a natural reversion to the profits from those enterprises in the sense of the above range of ideas. It is the real foster-mother [die eigentliche Nährmutter] of the same [i.e., the profits]; if the university did not exist, nothing would exist of these enterprises. For already the very beginning of them was demonstrably dependent on the rise that natural scientific research took precisely here in Jena forty years ago: the direct consequence of stimulation that the personal bearers of this rise gave at the time. And the university's decisive influence in the indirect or direct advancement can again be very clearly recognized in each decisive stage of further development through the impulses radiating from it [i.e., the university].

Somuch concerning the first motive for my offer.

The second motive stands on the ground of purely factual considerations and practical goals. I believe that I may say (and am strengthened in this opinion by the view of an impartial judge) that during the past forty years – during the latter half of this period under essential participation from my side – something has been created in Jena whose maintenance, further development, and continued security seems to be something that is a matter of public interest, due as well to its – to be sure only local – economic importance as also to its specific value that are attributable to the local institutions with respect to certain scientific and technical matters.

The two institutes involved here, the optical workshop [i.e., the Zeiss Werke] and the glass foundry [i.e., Schott & Co.], have made Jena into a center of the so-called scientific (i.e., for the needs of the sciences) working industry.[51]

For these reasons, in May 1889 Abbe created the Carl-Zeiss-Stiftung (Carl Zeiss Endowment) in honor of his late partner. Apart from providing for the future financial well-being of his wife and two daughters, Abbe arranged to bequeathe all his equity in both the Zeiss Werke and Schott & Co. to the Stiftung. In the meantime, he also provided an annual 20,000 marks to the Stiftung along with the possibility of receiving extraordinary funds.[52]

Yet Abbe soon felt dissatisfied with this arrangement: he wanted to see the

Stiftung take effect before his death.[53] Therefore, in 1896, the year in which the
Zeiss Werke celebrated its first half-century of existence, he arranged for the
Stiftung to purchase both his own and Roderich Zeiss's equity in the Werke and
in Schott & Co. (In 1903, Schott transferred his own remaining fifty percent
equity to the Stiftung.) By 1895/96, the Stiftung had a net worth of 2,000,000
marks, including assets of over 3,000,000 marks (of which 2.3 million was
equity in the Zeiss Werke and Schott & Co.) and debt of some 1,000,000 marks.
During the first six years of its existence (1889–95), the Stiftung gave a total of
255,000 marks (an average of 42,500 marks per year) to the University of
Jena.[54] Indeed, the support of teaching and research in natural science became
Abbe's first and principal interest.[55] Since the small state of Thuringia simply
could not support a first-rate university, the Stiftung stepped into the breach by
providing, on the one hand, regular, planned annual financial support to the
university (the 20,000 marks annually) so as to maintain and improve
institutes, equipment, apparatus, and faculty salaries; and, on the other,
extraordinary grants for special, one-time purposes. Particularly in its early
years, the Stiftung sought above all to support the institutes for mathematics,
physics, astronomy, mineralogy, and hygiene. (In 1890 alone, the Stiftung gave
an extraordinary gift of 70,000 marks for the construction of the astronomical
observatory.) And finally, the Stiftung and Otto Schott provided support for the
establishment of an Institute for Technical Physics and an Institute for
Technical Chemistry, institutes that have some special relevance for present
purposes.[56]

PART TWO: THE ULTRAMICROSCOPE

4. RICHARD ZSIGMONDY AND COLLOID CHEMISTRY
AT THE TURN OF THE CENTURY

Both the Zeiss Werke's and Schott & Co.'s extremely strong spirit of research
and development as well as their (especially the Werke's) financial ability to
realize that spirit materially meant that by the turn of the century both firms
stood as proven examples of the fruitful interaction of science and technology.
The Werke advanced and used physical knowledge; Schott & Co. advanced and
used chemical knowledge; and both firms made enormous profits and experienced unprecedented and unmatched growth as a consequence of their heavy
investments in research and development.

In the late 1890s, early 1900s, this corporate cultural environment of strong
support for research and development at the two sister firms found a new
subject for support in the "hot" young field of colloid chemistry, a field that
had, in principle, important potential benefits for the glass and optical
industries, among others. The Zeiss Werke was now ready to give material
support (on which more presently) to creating a new type of optical instrument
– the ultramicroscope – that would soon impart a new dynamic to the study of
colloidal phenomena. As Abbe neared the end of his days – he was sixty years

old in 1900, in poor health, and would die in 1905 – his firm stood ready to try to create an unprecedentedly powerful microscope, one that could detect objects even though they lay beyond the range of microscope resolving power that he had first theorized about in the early 1870s and one that would be surpassed only with the advent of the electron microscope in the early 1930s.

The catalyst who led the way towards uniting the support of the Werke with the needs of colloid research was Richard Zsigmondy (1865–1929). Zsigmondy's co-invention of an important scientific instrument perhaps owed something to his father, a Viennese dentist who, among other things, had invented a number of surgical instruments and devices.[57] Although Zsigmondy had dedicated himself since high school to becoming a chemist, he also had a strong interest in conducting physical experiments. With his joint interest in physics and chemistry, and with the emergence of the new discipline of physical chemistry in the 1880s, Zsigmondy might well have become one of the new physical chemists. Yet he did not. Instead, he studied organic chemistry, first at the Technische Hochschule Vienna, then at the Technische Hochschule Munich, and finally at the University of Erlangen, where he received his doctorate in that subject in 1889. Yet Zsigmondy had as little interest in organic as in physical chemistry. His real interest lay in the old-fashioned and relatively neglected field of inorganic chemistry. His move into this field owed much to his crucial decision to spend a year (1891–92) as an assistant to the University of Berlin's August Kundt, one of the foremost experimental physicists of his day.[58] Zsigmondy later spelled out the importance of his year with Kundt:

Following certain suggestions that I received from Kundt, I became concerned with the preparation of luster colors in glass and porcelain. I made the interesting discovery that the residues, following combustion of these organic solutions, which remained after heating thin layers of porcelain and whose coloring was reducible in the first instance to finely divided gold (according to the oxide contained therein), assume different colors. Thus, gold with silicon dioxide gave red colors, gold with titanium dioxide always [gave] blue colors, while gold with bismuth oxide gave, depending on the temperature, red or blue colors. Chemical analogy for the color of finely divided gold was thus not determinative, and so other causes had to be sought after. This and several other results led me to colloid chemistry.[59]

Those results also led him, first, to the Technische Hochschule Graz, where in 1893 he habilitated and became a lecturer in glass technology, and then to Schott & Co. – he had apparently known Otto Schott since at least 1895[60] – where in 1897 he became a scientific researcher concentrating on producing and studying colored and opaque glass. Yet Zsigmondy remained at Schott & Co. for only about three years: he found the work there too technical and not sufficiently scientific. Though he left Schott & Co. in 1900, he did not leave Jena until 1907: he remained there as a private teacher (at which he was most successful) and in 1903 married the daughter of Wilhelm Müller, who was professor of anatomical pathology at the University of Jena. Thus by 1900 Zsigmondy had become a free, institutionally independent colloid chemist in Jena who had good connections to local industry and the university.

At the turn of the century colloid chemistry was a dynamic field whose potential importance, many of its practitioners maintained, ranged from

understanding the foundations of matter and life itself to advancing the welfare
of more than a few branches of industry. As Zsigmondy stated in his book *Zur
Erkenntnis der Kolloide. Über irreversible Hydrosole und Ultramikroskopie*
(1905), one of the early masterpieces of the relatively new discipline, the field
was "of interest not only to the physicist and the chemist, but to scientists in
general". Jerome Alexander, an American colloid chemist, chemical engineeer,
and Zsigmondy's English-language translator pointed not only to the general
interests of physicists, chemists, and biologists in colloid chemistry, but also to
that of specific industries:

As the far-reaching ramifications of colloid chemistry are better understood, its importance and the
applicability of its principles to a great variety of industrial problems, become more and more
evident. There might, for example, be mentioned agriculture, tanning, dyeing; rubber, cement,
ceramics; soaps, photography, sugar – in fact, almost every industry is directly or indirectly
involved. Professor Zsigmondy's work will, therefore, be of vital interest not only to scientists
concerned with theoretical questions, but also to chemists, engineers, and others controlling
technical processes. To physiological chemists and physicians it is indispensable.[61]

Later, in 1917, the cautious Zsigmondy detailed some of the potential
significance of colloids to understanding life, and for use in medicine, farming,
and such industries as rubber, silk, ceramics, glass, dyeing, cement, alcohol,
and pharmaceuticals.[62] The broad theoretical and technical interest that
colloid chemistry offered fit perfectly into the spirit and program of the Zeiss
Werke and Schott & Co., and doubtless constituted in part the latter's
motivation for hiring Zsigmondy and, to anticipate, the former's support of
his efforts to build the ultramicroscope.

Colloids were first discovered and named in 1861 by the English chemist
Thomas Graham.[63] Graham provided the initial definition of colloids, contrasting them to crystalloids: colloidal solutions, unlike crystalloid solutions,
could not, he said, diffuse through a membrane. In time, it became further
known that colloids also differed from suspensions: colloidal metal solutions,
in contrast to suspended pulverized substances, divided differently, changed
their constitution irreversibly, and freed all their energy when separated from
their surrounding medium. Furthermore, Graham and his successors came to
perceive a wide range of substances as colloids; these included, for example,
dextrin, starch, various gums, molybdic acid and oxide, many albumens,
glycogen, caramel, tannin, stannic acid, metal sulphids, gelatin, glass, and
various sulphids, oxides, and metals.[64] Notwithstanding the study of colloidal
phenomena during the last third of the century, *circa* 1900 physicists and
chemists had achieved no consensus on defining and classifying colloids.

During the 1890s, first as Kundt's assistant, then as an independent lecturer
at the Technische Hochschule Graz, and finally as a scientific researcher at
Schott & Co., Zsigmondy focused his interest on colloidal gold solutions. An
inextricable mix of technical and scientific considerations had led him into the
field and, eventually, to his greatest achievements. It was, he later wrote,
"several technical problems [that] provided the incentive to concern myself
with colloids". He continued:

I was dealing with gold orpiment glass and with certain ceramic dyes which were based in principle on extremely fine gold divisions. It seemed to me particularly remarkable that the effect of very closely allied chemical compounds could be fundamentally different in regard to the preparation and appearance of these dyes. This observation seemed to me all the stranger as compounds with completely opposite properties frequently produce similar effects on the dyes of extremely finely divided (i.e. colloidal) gold.

Convinced that "the purely chemical" approach to analyzing colloidal phenomena was inadequate, Zsigmondy turned instead to physical approaches.[65]

He concentrated his attention on purple of Cassius, a colloidal substance named after its seventeenth-century discoverer, Andreas Cassius. By the early nineteenth century one of Sweden's leading chemists, Jöns Jacob Berzelius, was maintaining that purple of Cassius was a compound, while others were maintaining that it was "a mixture of finely divided gold and stannic acid". Late in the century Zsigmondy sought to resolve these conflicting views by synthesizing purple of Cassius. He developed a rather complex method for producing red colloidal gold; then used it to synthesize purple of Cassius; and showed the latter indeed to be a mixture of gold and stannic acid particles. Berzelius was wrong – it seemed. For Zsigmondy also found that "the purple [of Cassius] also behaved exactly like a chemical compound as Berzelius had already pointed out". Zsigmondy concluded: "*A colloidal mixture may sometimes behave like a chemical compound and has frequently simulated one*". Moreover, he found that the red colloidal gold was a solution, and one that behaved like a colloid, i.e., that its (molecular) gold particles mixed homogeneously with their liquid solvent. His initial results and conclusions were not well received; several chemists claimed that the red colloidal gold was a suspension, not a solution, i.e., that its (molecular) gold particles were only suspended in their liquid solvent. Zsigmondy later rejoined that his critics had "repeated my experiments without taking the necessary care" – and hence had obtained false results.[66] His results put him in professional battle with several rivals, and he needed to convince the competition and to prove that he was right and win over the field.

To prove that red colloidal gold was indeed a solution, Zsigmondy sought to answer three sets of questions, only two of which are important for present purposes. The first concerned the size of gold particles: How big are they? Are they molecules or molecular aggregates? And how many atoms are contained in a single particle? The second concerned the demonstration of a luminous path known as the Faraday-Tyndall cone, which emerges from an effect, present in all colloidal solutions, whereby a beam of light is visibly scattered along its path as it moves through a system containing discontinuties, for example, those of colloidal particles. Is the Faraday–Tyndall cone, Zsigmondy asked, an essential, or merely an accidental, characteristic of a group of gold particles in solution?[67] He apparently expected that particle size and display of a Faraday–Tyndall cone would, in general, indicate as to whether or not such particles constituted a colloidal solution or a suspension.

In trying to answer these two sets of questions, particularly the second, Zsigmondy found that even the best ordinary light microscope was inadequate for his purposes. He needed, instead, a microscope system that produced a brighter light cone than was then available. A preliminary set of experiments showed him that even extremely small particles that dispersed light

could be made individually perceptible by a microscopic examination of the light-cone; for if the small particles reflected enough sunlight, even if their size was below the limit of microscopic resolvability, they would ... be individually perceptible under the microscope and act to a certain extent as fragments of such light slits.[68]

In 1899 and early 1900, Zsigmondy conducted further preliminary observations and calculations that suggested the possibility of perceiving such particle images. Through the use of high magnification he observed

the presence of thousands of shining gold particles, whose size, as was shown by a rough calculation based upon the distance of the particles from each other and the amount of gold present, must have been less than the wave-length of light. With ordinary illumination, even with the best objectives, they were not perceptible.[69]

In other words, although Abbe's predicted limit of microscopic resolving power had been reached, Zsigmondy's preliminary research had nonetheless "convinced [him] that by the use of better objectives the individual particles could also be seen in Au_{60} and similar fluids, and I determined to attempt, with the assistance of a specialist, to render such particles visible also".[70] Zsigmondy now needed the resources of the Zeiss Werke to turn his convictions into proof – to show the images of the colloidal particles – that would convince others as well.

5. THE ULTRAMICROSCOPE: INVENTION, OPERATION, AND FIRST RESULTS

The specialist in question was Henry Siedentopf, a physicist who had studied first at the University of Leipzig and then at Göttingen with Woldemar Voigt, where in 1896 he received his doctorate for a dissertation entitled "Über die Capillaritätsconstanten geschmolzener Metalle" ("On the Capillarity Constants of Molten Metals") and who, three years later, joined the Zeiss Werke as a research physicist.[71] The Werke's hiring of Siedentopf fit a pattern: between 1898 and 1902 it hired no fewer than seventeen new scientific associates, most of whom were young, highly qualified physicists. By 1903, there were twenty-four scientific associates working at Zeiss. In fact, by the turn of the century the firm was run by physicists, several of whom also held faculty positions at the University of Jena.[72]

These physicists brought their academic culture and mores into the plant. Already in the winter of 1895–96, Abbe had introduced a weekly journal club that was attended by his scientific associates and, after 1898, by physicists and mathematicians from the Univeristy of Jena. At club meetings, his associates reported on the latest results published in optics journals and on their own research progress.[73] Abbe was now more than ever seeking to strengthen the Werke's science-base, no doubt in part because his own powers were in decline.

By 1900, Siedentopf belonged to the inner circle of optical physicists and technicians surrounding Abbe, who, until his retirement in March 1903, remained *au fait* with scientific developments at Zeiss.[74] More to the point: Zsigmondy's and Siedentopf's research and development of the ultramicroscope at the Werke only began following a discussion of it with Abbe.[75] In sum, Zsigmondy's prior employment with Schott and Co., his continued presence in Jena as a private science instructor, Siedentopf's expertise in the physics of metals, and, above all, the potential relevance of understanding and applying colloidal substances to advance various aspects of glassmaking and ceramics, made it only natural that Zsigmondy now turned to Siedentopf and the Werke for assistance.

For a year and a half – from April 1901 to October 1902 – Siedentopf and Zsigmondy worked individually and together on their project: they theorized about, planned, constructed, tested, and used the resources of the Zeiss Werke to create the ultramicroscope.[76] They did not hesitate to acknowledge the Werke's support. As they wrote in the conclusion to their classic, 1903 presentation of the ultramicroscope:

It is the authors' pleasant duty to express their best thanks to the firm of Carl Zeiss in Jena for so liberally making available the means to carry out this investigation.[77]

Zsigmondy made the same point again two years later, this time even more explicitly. He wrote:

We received splendid assistance in our work from the numerous means placed at our disposal by the firm of Zeiss. We were thus enabled to work with the best [optical] objectives, and complete the mechanical development of the apparatus in a short time. Notwithstanding this, a year and a half of study was necessary to bring the apparatus to its present state of perfection.[78]

The Werke's material support was thus essential to Zsigmondy's and Siedentopf's project to construct a new type of microscope, one that enabled them to "see" particles below the limit of optical resolution that Abbe had predicted thirty years previously.

The division of labor between the two men was clear and straightforward. The original idea of an ultramicroscope and its purpose was Zsigmondy's alone. In 1909, Zsigmondy testily (if only privately) criticized Svante Arrhenius, the Swedish physical chemist, for his slip in mistakenly referring to "Siedentopf's ultramicroscope":

The complete suppression of my name has to be all the more insulting to me since the apparatus originated solely at my instigation and under my constant collaboration. I devoted a great deal of time to this, in the conscious striving to bring to completion, in partnership with a specialist, an important method for my work and also for colloid research, among other things.[79]

Siedentopf, for his part, constructed and improved upon the ultramicroscope itself, while Zsigmondy alone tested it with his colloidal preparations and developed the method for determining the size of ultramicroscopic particles. In addition, the two men together developed the theory of rendering the ultramicroscopic particles visible and calculated the limits of visibility.[80]

The ultimate object of Siedentopf and Zsigmondy's first ultramicroscopic investigation was the rendering visible ("Sichtbarmachung") and determination of the order of magnitude of the colloidal substance gold ruby glass. Previous researchers knew nothing about the properties of such substances since their size lay below the half-wavelength of light and, hence, it was assumed that they could not be seen. Siedentopf and Zsigmondy therefore sought to develop a method of "visualizing" such microscopic substances. Their eighteen-month investigation allowed them nearly to reach the limits of visibility. They wrote of their investigation's general results and implications:

With its [i.e., the method's] help, we were able to make visible individual gold particles whose sizes are not very far from molecular dimensions. We emphasize further that our method is capable of a more general application, and that we ourselves have already, with some success, extended the investigation to the study of colloidal solutions and to more opaque media. However, since such investigations are of a more chemical or medical interest, we must limit ourselves here [i.e., in the *Annalen der Physik*] first to discussing the optical conditions of the method and then to describing it [i.e., the method] itself.[81]

Accordingly, Siedentopf and Zsigmondy divided their paper into three major sections: 1) "On a Method for Rendering Ultramicroscopic Particles Visible", wherein they discussed the ultramicroscope, its theoretical principles, and its methods of use; 2) "On a Method for Determining the Size of Ultramicroscopic Particles"; and 3) "Relations between Color and Particle Size in Gold Ruby Glasses".

In section 1, then, Siedentopf and Zsigmondy presented their method for rendering the particles visible. They argued that thin-sectioned gold ruby glasses behaved, in principle, like colored bacterial preparations; hence, they should have a heterogeneous nature. Yet they wondered how such a nature could be displayed given "that the size of the particles to be rendered visible would presumably lie not insignificantly below that which is still accessible to the best microscopes". The principal insight to visualizing such particles lay, they argued, in the realization that the diffraction disks (*Beugungsscheibchen*) of self-illuminating particles of high radiating power could be seen microscopically "even if the particles are smaller than a half-wavelength of visible light". In other words, and to be somewhat more precise, if the size of the particle was less than half a wavelength, then the image would be (only) a diffraction disk. The only condition here was that the product of the light intensity and the sine squared of the angle of illumination had to be greater than the eye's lower limit of sensitivity. Hence the surprising conclusion that the limit sought for rendering such particles visible would be "of a far small order of magnitude ... than the limit determined by Abbe and Helmholtz". Now, since such particles are not self-illuminating, light (e.g., arc light or sun light) had naturally to be shone on them. Such illuminating light rays are, however, much more intense than the diffracted rays. "Thus, in order to render small particles visible through their diffraction cone, it became a *principal requirement that the illumination must be so arranged that none of the illuminating rays*

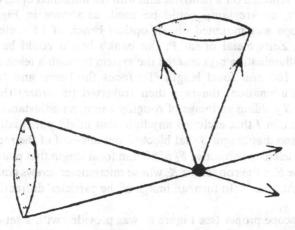

Figure 3. Schematic diagram showing the vertically oriented diffraction cone of the microscope objective perpendicular to the horizontally oriented illumination cone of the condensor. Source: Siedentopf and Zsigmondy, "Über Sichtbarmachung", 4 (see note 76).

are contained in the diffraction cone used to render [the particles] visible." This could be done through traditional dark-field illumination so long as "the axis of the illuminating cone stands perpendicular to the axis of the diffraction cone" used to render the particles visible. (See Figure 3.) Here they brought into their method the old idea of the Faraday–Tyndall cone: the particles were made visual by diffraction of the cone's light rays. As long as the cones' axes are exactly perpendicular to one another, then the light from the illuminating cone cannot reach the eye while only that from the cone diffracted upwards into the microscope will. Herein lay the refinement of traditional dark-field illumination. This joining together of an older idea – that of the Faraday–Tyndall cone – with an advanced but nonetheless traditional instrument – a (Zeiss) microscope – was the key innovation that led to the ultramicroscope and, as we shall see, its breakthrough results for colloid chemistry and consequences for matter theory.[82]

As the condensor, they used a Zeiss microscope system with a numerical aperture of only 0.3 but with a large opening for the illumination. The adjustment of the illuminating rays was "of decisive importance". To achieve the proper adjustment – that is, to prevent the overshadowing of the ultra-microscopic particles' diffraction disks – they thus used a precision slit to exclude all light scattered from particles larger than those of ultramicroscopic dimensions. After passing through the slit, the condensor then focused the light onto the ultramicroscopic particles.[83]

Figure 4 is a schematic diagram of Siedentopf and Zsigmondy's first ultramicroscope, the so-called slit-ultramicroscope. For illumination, they used sunlight reflected off a heliostat and into the darkened observation room. (Alternatively, an arc lamp could be used, as shown in Figure 5.) The ultramicroscope was mounted on an optical bench of 1.5 meters in length supporting a Zeiss metal prism P; the bank's height could be adjusted as needed. The illuminating rays entered the system through a telescope objective F_1 of about 100 mm focal length. To focus the beam and to avoid any extraneous illumination, the rays then traversed (in order) the adjustable precision slit S yielding an image of roughly 1 mm; an adjustable polarizer N; an iris diaphragm J that excluded any light that might potentially enter from the side; an iron diaphragm B that blocked out one-half of the radiation cone; and a second telescope objective F_2 of 80 mm focal length that placed the image onto the plane E of the condensor K, whose micrometer screws could adjust the size of the light cone.[84] In turn, an image of the particles' diffraction disks was formed.[85]

The microscope proper (see Figure 6) was provided with a set of adjustable screws that allowed appropriate positioning of the preparation under investigation so that the illuminating rays transmitted through the condensor could be brought into the illumination cone. For studying liquid preparations a special holder with a water-jet installation was used (see Figure 7); it was placed directly under the immersion objective or, more conveniently, a small glass apparatus (see Figures 8 and 9). Siedentopf and Zsigmondy pointed out that much care had to be taken in getting "pure" liquid (or solid) preparations, and that it took some practice in learning to manipulate the preparation so that its rays transmitted into the light cone. *En passant*, they added that the technique was "especially appropriate for the study of the so-called Brownian motion in liquids".[86]

The suspended or mobile particles produced diffraction disks that were seen within the cone; the disks' size, color, and intensity depended on the preparation in question. Figure 10 shows disks in four different polarization planes, with the respective diffraction disks shown (schematically) below each circle. Figure 10 shows the images of the disks that Siedentopf and Zsigmondy saw. They argued that for the human eye the theoretical limits of rendering ultramicroscopic particles visible was about 36 $(\mu\mu)^2$ or, in modern notation, 36×10^{-12} pm^2, and that their observations came, in terms of order of magnitude, "very close" to that limit.[87]

In section 2, Zsigmondy explained the method for determining the size of ultramicroscopic particles. The length of such particles, he found, ranged between 0.006 and 0.25 µm. Yet the size of the diffraction disk did not permit any conclusions about the size of the particle itself. To determine the size, Zsigmondy chose (by way of example) gold ruby glass particles. Assuming A is the concentration of a cubic millimeter of gold metal glass in milligrams, n the number of gold particles in a cubic millimeter, and s the specific weight of the gold, then the length of a side of a gold particle (assuming it to be a cube) was

$$l = \sqrt[3]{\frac{A}{sn}} .$$

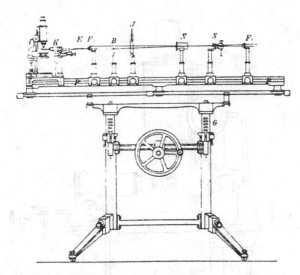

Figure 4. Schematic diagram of Siedentopf and Zsigmondy's first ultramicroscope, the so-called slit ultramicroscope. Source: Seidentopf and Zsigmondy, "Über Sichtbarmachung", 7 (see note 76),

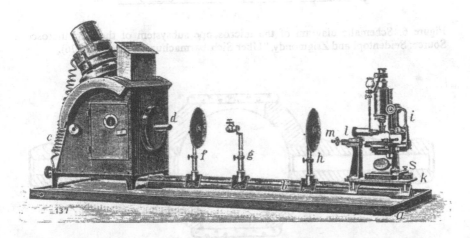

Figure 5. A slit ultramicroscope for examining fluids. (On the left is an arc lamp *c*). Source: *Description of Optical Appliances to Facilitate Visual Perception of Ultramicroscopic Particles. Catalogue.* Carl Zeiss, Jena, 1904, 3.

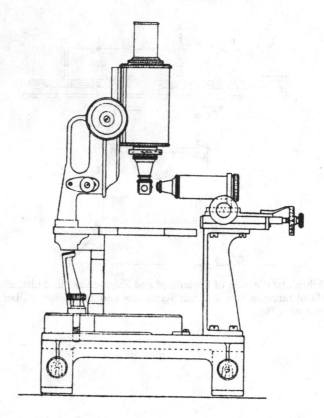

Figure 6. Schematic diagram of the microscope subsystem of the ultramicroscope. Source: Seidentopf and Zsigmondy, "Über Sichtbarmachung", 8 (see note 76).

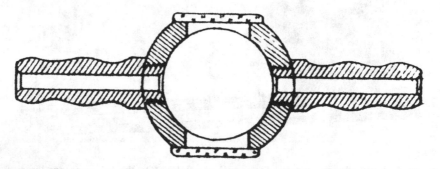

Figure 7. Schematic diagram of the holder for studying liquid preparations with the ultramicroscope. Source: Seidentopf and Zsigmondy, "Über Sichtbarmachung", 9 (see note 76).

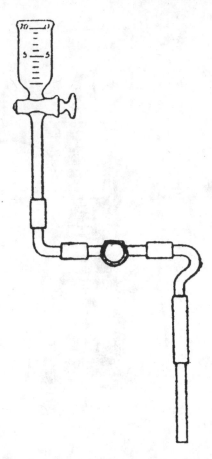

Figure 8. Schematic diagram of glass apparatus used to transmit liquid preparations in the ultramicroscope. Source: Seidentopf and Zsigmondy, "Über Sichtbarmachung", 9 (see note 76).

Since A and s were known, the problem became one of determining the number of particles n in a cubic millimeter. Zsigmondy accomplished this "difficult and laborious work" literally by counting the particles in a highly concentrated area (as shown in Figures 11a and 11b) and then, from that count, calculating the amount for a cubic millimeter. Such counts, he warned, had to be done repeatedly and in different areas. He found that particles ranged in size between 6×10^{-6} and 2.5×10^{-4} mm, or between 2.16×10^{-16} and 1.5×10^{-11} cm^2. In a long discussion of potential sources and limits of error he argued that his figures were quite precise.[88]

Finally, in section 3 Zsigmondy presented his results of the initial use of the ultramicroscope, that concerning the relationship between the color and particle size of gold ruby glasses. Here methodological considerations proved

Figure 9. A microscope with tubular apparatus for easily introducing and disposing of fluids containing ultramicroscopic particles. Source: Zsigmondy, *Colloids and the Ultramicroscope*, 107 (see note 61).

essential. The size, color, and other optical properties of gold ruby glasses depended not only on glass composition but also on the heat-treatment process it underwent during the production phase.[89] Hence in discussing size, color, and so on the ultramicroscopist had to be extraordinarily precise in specifying which type of glass had been observed. Zsigmondy emphasized, moreover, that he counted only between fifteen and thirty gold particles in a limited area and then repeated this process in neighboring areas; all told, his characteristics for gold particles were based on 100 to 200 such particles in a given region. He found that glass size ranged between 3.9 and 6.9 pm on the lower end to 487 and 791 pm on the upper. He also found that in general color did not seem to be related to size. Finally, he pointed out that it was well known that it is possible to produce considerably smaller divisions of gold ruby glass (much smaller

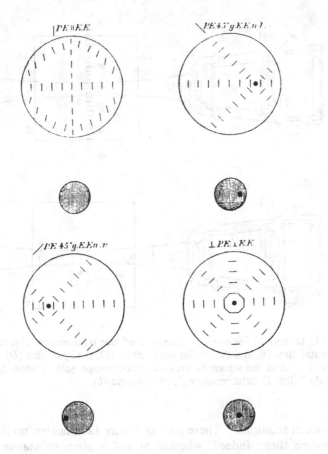

Figure 10. Schematic diagram of four diffraction disks in four different polarization planes. Source: Seidentopf and Zsigmondy, "Über Sichtbarmachung", 12 (see note 76).

than 3–6 pm), especially in the liquid state. Apart from a few brief preliminary remarks, however, he reserved discussion of this fundamental subject of colloidal phenomena for a future study.[90]

In point of fact, Zsigmondy already had such studies underway, and in his book of 1905, *Zur Erkenntnis der Kolloide*, he related the history and results of colloid chemistry to date, explained the ultramicroscope's principles and use, and reported on new findings by himself and others. There he summed up his (and others') findings to date on colloidal gold particles by noting that, since particle-size measurements were also a function of a given substance's specific gravity, the internal friction of the fluid, and, in some cases, the particles' electrical charges, no fixed upper and lower limits to size could be given. Instead, he argued that crystalloids, colloids, and suspensions belonged to a

a

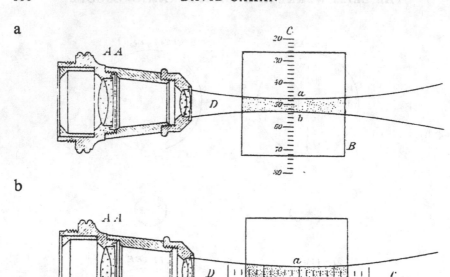

b

Figure 11. (a and b). Schematic diagram of the microscope objective (*AA*), the experimental glass (*B*) and the ocular micrometer (*C*), the light cone (*D*), and the width of the light cone at the strongest area of concentration (*ab*). Source: Seidentopf and Zsigmondy, "Über Sichtbarmachung", 18 (see note 76).

continuum of solutions.[91] There was, as Figure 12 indicates, no sharp dividing line between them; indeed, whether or not a given substance is a colloid depended on its physical state and surrounding medium. Still, particle size was important for determining where to classify a substance.

Along with reporting numerous other experimental findings in his book, Zsigmondy now also emphasized a point that he and Siedentopf had scarcely mentioned in their joint study, a point that was to be of central importance for physicists and chemists in the next few years: ultramicroscopic particles displayed Brownian motion. Zsigmondy here gave a dramatic, if not romantic, description of the Brownian motion that he had observed:

> The small gold particles no longer float, they move – and that with astonishing rapidity. A swarm of dancing gnats in a sunbeam will give one an idea of the motion of the gold particles in the hydrosol of gold! They hop, dance, jump, dash together, and fly away from each other, so that it is difficult in the whirl to get one's bearings ....
>
> Sluggish and slow in comparison is the analogous Brownian movement of the larger gold particles in the fluid, which are the transition forms to ordinary gold that settles.

Zsigmondy found that the colloidal gold particles not only showed Brownian motion but that the smaller the particle, the faster its rate of motion. Although

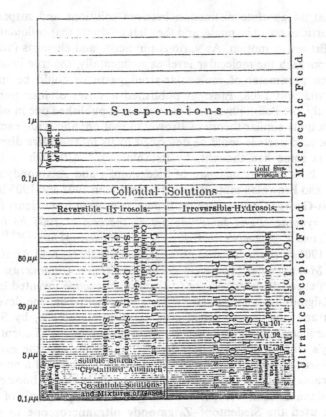

Figure 12. Diagram showing the classification of colloidal solutions and suspensions according to particle size. Source: Zsigmondy, *Colloids and the Ultramicroscope*, facing p. 26 (see note 61).

he did not measure any rates of motion, he recognized that colloidal particles characteristically experienced Brownian motion. As to the causes of Brownian motion, he could only call for further investigation.[92]

6. CONCLUSION

Such investigations came virtually immediately and, as one indicator of the ultramicroscope's historical importance, they merit brief notice here.[93]

Brownian motion (and especially its causes) puzzled not only Zsigmondy; for three-quarters of a century, ever since Robert Brown first discovered it in 1828, Brownian motion had puzzled a variety of botanists, chemists, and physicists. Not the least surprising thing about it was that, in part as a consequence of Zsigmondy's creation and initial use of the ultramicroscope, such motion became intertwined with the kinetic theory of matter and the proof of the existence of atoms. Zsigmondy's ultramicroscopic studies showed, as noted already, that no absolute distinction between solutions and suspensions of colloidal particles could be made, and they left no doubt that colloidal particles displayed Brownian motion. As various physicists and chemists (and others) began to approach the molecular level experimentally, the true issue became clear: which parameter of molecular-sized particles could be measured? Between 1905 and 1908, Einstein published a series of four papers and, independently, the Polish physicist Smoluchowski published one in which each developed a quite similar theory of Brownian motion; above all, each showed that a suspended particle could be described by its mean-square displacement (and not, as some had thought, by its velocity).[94] Although in his initial publication Einstein was unaware of Siedentopf's and Zsigmondy's findings on colloids and Brownian motion, sometime between 1902 and 1905 he learned about Louis-George Gouy's work on Brownian motion from Henri Poincaré's *Science et hypothèse, and sometime before the end of 1905 he learned of Siedentopf's and Zsigmondy's work directly from Siedentopf himself.*[95]

Then in 1906, a young Swedish chemist, The Svedberg, read Aimé Cotton's and Henri Mouton's *Les Ultramicroscopes et les Objets Ultramicroscopiques.*[96] The French chemists described the (slit) ultramicroscope invented by Siedentopf and Zsigmondy as well as two other (and new) types, one developed by themselves and one, the so-called co-axial ultramicroscope, by Siedentopf alone.[97] Svedberg became enthused by the possibilities that Siedentopf's and Zsigmondy's slit ultramicroscope offered towards proving the existence of atoms – whose existence he never doubted – and he obtained a version of it. He also became equally enthused and inspired by Zsigmondy's book of 1905 on colloids and the ultramicroscope.[98] Working at first independently of Einstein's theory, he used the Siedentopf–Zsigmondy ultramicroscope to study the Brownian motion of colloidal metal particles, seeking to show that such motion was due to molecular thermal motion.[99] Shortly after publishing his results and claims, he learned of Einstein's theory, which he then declared his experimental work to have confirmed.[100] As Einstein quickly pointed out, however, Svedberg had misunderstood his theory. For Svedberg claimed to have measured the particle's velocity, something that Einstein (and other theorists) showed was quite impossible to do.[101] Although Svedberg's name would long continue to be cited as having adduced experimental proof of Einstein's and Smoluchowski's theory of Brownian motion, it was, in point of fact, Jean Perrin who really provided such proof. Once Perrin realized the combined importance of Zsigmondy's ultramicroscope, colloid chemistry, and Brownian motion for proving the molecular hypothesis, he produced (in 1908 and 1909) solid experimental evidence for Einstein's and Smoluchowski's theory. Not the least important facet of Perrin's work on Brownian motion was that it led to a new and better value for Avogadro's number, which itself provided a crucial point in favor of the argument for the existence of atoms.[102]

Zsigmondy's determination of the size of various colloidal particles and his (and Siedentopf's) creation of the ultramicroscope as a revolutionary means for viewing colloidal phenomena quickly became widely accepted. In 1907,

Wilhelm Ostwald, the physical chemist and editor of the *Zeitschrift für physikalische Chemie, Stöchiometrie, und Verwandtschaftslehre*, reviewed Zsigmondy's *Zur Erkenntnis der Kolloide*, and wrote of the ultramiscroscope and its implications:

Through the process developed by the author [i.e., Zsigmondy] and Siedentopf for rendering particles whose size lies below that of a wavelength of visible light, a large, previously unknown field of facts has been opened up for research. The incorporation of this new material into the old stock of science is proving to be extremely difficult work. Many an old boundary that has been held for certain is now being affected.[103]

Ostwald, always partial to interdisciplinarity and always quick to generalize, saw the instrument's great potential. Ironically, one of its principal uses was to give the *coup de grâce* to Ostwald's pet philosophy of matter: energeticism. A year later, in 1908, Ostwald suddenly acknowledged the reality of atoms.[104]

As for Zsigmondy, his colloidal studies and invention of the ultramicroscope soon brought him more than just intellectual rewards. In 1907, he was appointed extraordinary professor of inorganic chemistry and director of the Institute for Inorganic Chemistry at the University of Göttingen; in 1919, he was promoted to ordinary professor. Starting in 1913, furthermore, he received the first of eleven nominations for the Nobel Prize in Chemistry. Daniel Vorländer, professor of chemistry at the University of Halle, who nominated him that year for the chemistry prize, said that he found Zsigmondy's work to be more important than that of such chemical luminaries as Walther Nernst, Richard Willstätter, Hermann Staudinger, and Alfred Werner. Vorländer wrote:

The name Zsigmondy is intimately tied to the newest development in chemistry, to inorganic as well as to organic and to theoretical chemistry. This development can be expressed in a few words: that the recognition of the importance of *colloids* permeates and fructifies the entire field of chemistry. Zsigmondy gave not only the first stimulus to this new development but has also led, right down to the present, colloid chemistry to its current general recognition.

Vorländer continued:

Zsigmondy's work on colloidal gold solutions and gold glasses gave the impulse to the discovery [*Entdeckung*] of the ultramicroscope in the year 1902–1903. ... He constructed the first, if rather primitive, ultramicroscope, and gave the firm of Carl Zeiss in Jena and its outstanding optician, Siedentopf, the inducement to construct an apparatus that is today in wide use. ... With the discovery of the ultramicroscope a new era in colloid chemistry and in chemistry in general begins. ... The significance of Zsigmondy's work is seen not only in the field of chemistry right down to the present; physiology, biology, and hygiene have also recorded much progress by using the ultramicroscope. The direct proofs for the existence of molecules, for the parallels between the Brown–Zsigmondyian movement of submicrons, and for the movement of molecules in a gaseous state – work that in the past year was done by The Svedberg – have become, on the basis of ultramicroscopic measurements and observations, of inestimable importance for physics and chemistry.[105]

A year later Georg Wiegner, professor of chemistry at the Eidgenössische Technische Hochschule in Zurich, also nominated Zsigmondy for the chemistry prize. He noted that colloid chemistry and the chemistry of radioactivity "have in recent years contributed to general chemistry in outstanding and original ways by intensely deepening our theoretical knowledge and by applying this knowledge in practice". In particular, he pointed to colloid chemistry's applications to botany, physiology, medicine, photography, rubber technology, dyes, tanning, the glue industry, agricultural chemistry, soil science, and so on. He continued:

It is recognized well enough that the development of an experimental science is frequently connected with the invention of a new apparatus that expands our sensory means beyond the current level; indeed, that in general an experimental science frequently proceeds from such an invention. The development of modern colloid chemistry is, in this regard, typical. The rather sparse literature on colloid chemistry and the scattered interest in its problems until the year 1903 received a powerful stimulus with the invention of the ultramicroscope. It allowed penetration into a previously inaccessible region that, as it turned out in the course of research, was precisely characteristic for the colloid state. The pathbreaker in this new region and the inventor of the slit ultramicroscope – that alone was well suited for chemical investigations – was, along with Siedentopf, Richard Zsigmondy, who, after he had previously become involved with a full series of studies on the chemical and physical properties of glasses, at first cultivated the chemical side of the new discovery [*Entdeckung*]. The discovery proved decisive concerning the old (since Faraday) and much discussed question about the discontinuity of colloidal systems; beyond that, however, it above all gave impetus to a fundamental examination of a large field of science that had previously been rather uncultivated. All at once a taxonomy and a completely new, original exploration was possible. In the field of the newly created ultramicroscopy Zsigmondy was, until the last few years, independently and successfully active. ...

Not only were the optical methods of colloid research in part created or in part expanded by Zsigmondy; there is hardly any area of colloid chemical, experimental research which has not gained profit through the improvement of the apparatus. ...[106]

After the First World War, in 1919, nominations for Zsigmondy to receive the prize recommenced. Rudolf Wegscheider of Vienna nominated him that year;[107] Wilhelm Ostwald nominated him (and Siedentopf) for 1921, 1923, and 1924, characterizing them "as founders of ultramicroscopy and, thereby, of an important branch of colloid chemistry";[108] Hugo Kruyt of Utrecht also nominated Zsigmondy for 1921;[109] Wiegner again nominated him, this time for 1922;[110] Karl Albert Vesterberg of Stockholm nominated him for chemistry for 1923;[111] Arthur Kötz of Göttingen nominated him for 1925;[112] and Svedberg, Georg Meyer of Freiburg, and Viktor Rothmund of Prague all nominated him for 1926.[113] Zsigmondy finally received the Nobel Prize in Chemistry for 1925 (reserved for 1926) "for proving the heterogeneous nature of colloid solutions and for the methods he used which have laid the foundations of modern colloid chemistry".[114]

Moreover, the use of the ultramicroscope as an instrument for helping to provide the experimental foundations of the kinetic theory (and hence of atomism), for understanding Brownian motion, and for advancing colloid chemistry meant that it played an indirect role in the awarding of the Nobel Prize to Einstein (physics, 1921), Perrin (physics, 1926), and Svedberg (chemistry, 1926). As for Siedentopf, in 1907 he was appointed director of the Zeiss Werke's microscope laboratory, a position he maintained until his retirement in 1938.[115]

This essay has sought to understand the scientific, technological, industrial, and economic context in which the ultramicroscope was created and how that context helped shape scientific knowledge, which in turn helped reshape the context from which it emerged. It has sought to do so by studying events on the local, national, and international levels as well as within several scientific disciplines and industries. As the essay has implicitly sought to show, the history of the creation of the ultramicroscope, perhaps like the history of all instruments, can provide a useful vehicle for understanding not only the creation of knowledge but also the relationship between that creation and its context.

If the ultramicroscope was an invention – as, for example, the various testimonies nominating Zsigmondy (and Siedentopf) for the Nobel Prize unquestioningly asserted – it was not a *de novo* invention. Indeed, in some ways it might better be called an innovation. Siedentopf and Zsigmondy drew on accumulated knowledge about microscope theory and the practical construction of microscopes, and they employed previous innovations in optical glass technology that Zeiss, Abbe, Schott, and their co-workers had developed in the prior half-century. They used and were quite conscious of the limits of microscopic resolution as shown by Abbe and Helmholtz in the early 1870s; at the same time, they realized that such limits did not necessarily prevent them from "seeing" (i.e., imaging) colloidal particles whose size lay below the limit of resolution. Zsigmondy and Siedentopf's understanding of optical theory owed much to Abbe and Helmholtz. As Zsigmondy wrote:

> It must be here noted that the limits of microscopic resolution (limit of visible 'separation') determined by Abbe and Helmholtz is often confounded with the limit of 'visibility'. That isolated particles whose diameter is a fraction of a wavelength of light can still be seen, is expressly stated by Abbe himself. He writes in this connection: 'Such objects can be seen, no matter how small they may be; it is only a question of the contrast of the light effect, good definition of the objective, and sensitiveness of the retina.' As may be seen, the results of our investigation are not at all, as has been often thought, in conflict with the statements of Abbe.[116]

Although neither Siedentopf and Zsigmondy nor anyone else who subsequently worked with an ultramicroscope ever, of course, saw ultramicroscopic colloidal particles directly, they did see the optical effects of such particles, and that was enough to prove the particles' existence and, to a certain extent, give information as to their size, color, and motion.[117] Herein lay part of Siedentopf's and, especially, Zsigmondy's originality.

Moreover, the key insight that led to the ultramicroscope was, as we have seen, Zsigmondy's creative use of the older idea of the Faraday–Tyndall light cone. The conjoining of such a cone (and a Nicol prism) to an otherwise traditional microscopic set-up constituted the innovation (invention) that became the ultramicroscope. Siedentopf and Zsigmondy's ultramicroscope was thus the material embodiment of Abbe's and Helmholtz's old ideas about the centrality and use of diffraction in resolving objects as well as that of the scattering of the Faraday–Tyndall cone. In addition, Siedentopf and Zsigmondy used otherwise traditional if nonetheless advanced Zeiss components (e.g., the microscope objective) as the constituent elements of their ultramicroscope.

While the ultramicroscope embodied both standard optical theory and material goods, in its capacity as a new, more powerful instrument it also helped advance scientific knowledge. Soon after its announcement in 1903, it catalyzed and accelerated research in the field of colloid chemistry and quickly helped to lead, as we have also noted, to understanding of the causes of Brownian motion and the experimental proof of kinetic theory, and to the conviction among virtually all scientists that matter is ultimately constituted by atoms.

The Zeiss Werke had flourished principally by constantly seeking to prosecute scientific theory (above all, optics) as the *sine qua non* for future technological innovation in optical instrumentation; by employing physicists (and others) trained at Germany's universities and *Technische Hochschulen*; by improving understanding (and the processing) of raw materials, above all that of glass; and by introducing a strong division of labor that included the separate production of each part of an optical instrument and careful testing to ensure that each part, and the instrument as a whole, met pre-determined scientific and technical specifications and tolerances. In short, following the entrance of Abbe as consultant to the firm in 1866 the Zeiss Werke operated on the basis of an increasingly rationalized system of production, one in which scientific theory was constantly advanced and tested against technological performance and practice.

Abbe had thus created at the Werke an extremely favorable local culture in which science and industry flourished together, not least by associating highly skilled workers within one setting (i.e., an industrial plant).[118] Siedentopf and Zsigmondy, supported by an array of (other) workers and material resources, had the skills needed to create and use a new, highly sophisticated instrument like the ultramicroscope. An essential part of the Zeiss culture, perhaps its essence, was the creation of an atmosphere in which the social relations between skilled technicians and (more) theoretically oriented scientists could work to their mutual advantage, as well as to the advantage of the Werke as a business entity. The creation and use of the ultramicroscope is thus an(other) instance of how the history of theory is intimately and inexorably tied to the history of instrumentation.[119]

It was in this singular industrial atmosphere and within this historical framework that the ultramicroscope was (in part) conceived and (totally) produced. In the event, no other historical setting of the day provided the considerable financial and scientific resources and conditions needed to create the ultramicroscope. While the ultramicroscope was certainly the brainchild of the inorganic chemist Zsigmondy, who was then seeking to use it as part of a specific scientific research program, it is nonetheless also true that he, Abbe, and other senior scientists at the Werke recognized the ultramicroscope's potential application to a myriad of fields. Support for the ultramicroscope thus helped keep the Werke at the forefront of microscope research and development while also promising potential profits. The Werke provided a friendly, nurturing environment for research and development.

It did something else as well. The name behind the ultramicroscope's creation and first use was itself most useful in gaining trust for Siedentopf and Zsigmondy's instrument and results. During the previous quarter century the Werke had achieved impeccable credentials in practical microscopy and gained an authority amongst scientists in general such that any product issuing from its plant and with its name on it was immediately seen favorably. Those scientists who followed the ultramicroscope's creation and later participated in its further development and use were predisposed to accept results achieved with a Zeiss product. The very name "Zeiss" became an element in persuading others to accept the work of Siedentopf and Zsigmondy. Belief in Zsigmondy's results among members of the scientific community was thus greatly accelerated. By the same token, the name "Zeiss" also helped win new customers (other scientists) for the ultramicroscope.[120]

The motivations of the Zeiss Werke to support the creation of the ultramicroscope were thus multiple and ineluctably interrelated. There were, for one, market considerations, meaning above all those of the long-term sort. Part of the ethos of the Zeiss Werke was to encourage calculated risk-taking, to explore previously uninvestigated or incompletely investigated scientific and technical ideas and fields that might contain sufficient potential for technological change and industrial profit.[121] To do so meant to invest in new scientific instrumentation, which, as Nathan Rosenberg has (rightly) maintained, constitute the *capital goods* of scientific research.[122] The ultramicroscope represented a form of investment on the part of the Werke: it permitted the further pursuit of scientific research through the increased capacity to observe and measure older and new phenomena and, by advancing research in microscope development, it held forth the promise of creating a new sector in the microscope market and, with it, both short- and long-term profits. The potential market here included all those industries that produced or consumed colloidal materials.[123] The Werke's history, as we have seen, was one of constant and heavy investment in research and development in order to secure new markets and, hence, the company's long-term survival and financial well-being. By 1900, the Werke had become an enormously successful and wealthy organization, and so support for Siedentopf's and Zsigmondy's experimental microscope occurred more or less naturally. There was, for another, the instrumental challenge simply to build a new microscope; the firm had, of course, always sought to build new instrumentation. Constructing and developing the ultramicroscope was simply a challenge that had to be faced, and the firm's pride encouraged it not to let such challenges go unmet. Third and finally, there was the challenge to improve and test (microscope) theory in the guise of developing a means of extending the range of resolution beyond that enunciated long ago by Abbe. In short, the Werke's cultural ethos in the broadest sense – ranging from the pursuit of profits to that of pure science – led it to support the research and development of the ultramicroscope. There was no simple functional relationship at work here between research and development, on the one hand, and profit on the other. Instead, the ultramicroscope arose within a complex and dynamic scientific, technological, industrial, and economic context, one in which financial, instrumentational, and intellectual values all had a share.[124]

The history of the creation of the ultramicroscope in its multivalent context thus shows the difficulty (to say the least) of distinguishing between those contributions issuing from industry and those from universities and *Technische Hochschulen*. The ultramicroscope, like the Zeiss Werke itself, embodied the blurring of the science, technology, and industrial enterprises referred to at the outset of this essay. The ultramicroscope's inventors, Siedentopf and Zsigmondy, like Abbe and all the other scientific and technical leaders at the Werke from the late 1880s onward, were highly skilled individuals whose creative activities at Zeiss owed much to the advanced scientific training that they had received at various German universities and *Technische Hochschulen*. In the creation of the ultramicroscope, there was no unidirectional flow of goods (instruments), ideas, results, or even personnel from universities and *Technische Hochschulen* to industry (or vice versa). Throughout his career at Zeiss, Abbe recognized the mutual needs and problems of optics as practiced at universities, *Technische Hochschulen*, and in his own industrial firm. He saw and furthered a strongly complementary relationship and interaction between industry and academic science. Indeed, if one adds to this interaction the crucial financial support provided by the Prussian government during the mid-1880s to aid Schott's efforts at exploring potentially innovative types of glass, not to mention the Stiftung's support of scientific and technological education and research at the University of Jena, one sees just how blurred these enterprises could become.

Finally, the history of the creation of the ultramicroscope also provides an illustration of how well, in this instance at least, German scientists and technologists from industry and academic life could cooperate with one another to their mutual advantage. In his study of the work of physical chemists at General Electric from 1900 to 1915, George Wise has convincingly argued that there was a conflict of values between those in an academic and those in an industrial setting; he noted, moreover, that a number of recent studies of American science have arrived at similar results.[125] The present study, in contrast, points to a strong convergence of interests and values. To be sure, Zsigmondy did not much care for working in an industrial research laboratory, and it is also true that his association with the Zeiss Werke during the construction of the ultramicroscope was only of a short-term, ad-hoc nature. Yet he and Siedentopf worked extremely well together, and neither the Werke nor Zsigmondy ever found themselves in a conflict of values or interests. Indeed, quite the opposite was the case: the creation of the ultramicroscope, like the history of the Zeiss Werke as a whole, provides a striking example of what, at its best, the close, supportive, and mutually beneficial interaction of science, technology, and industry in Germany could achieve.[126]

University of Nebraska-Lincoln

NOTES

[1] For extensive recent bibliographies on the growing studies of scientific instrumentation see David Gooding, Trevor Pinch, and Simon Schaffer, eds., *The Uses of Experiment: Studies in the Natural Sciences* (Cambridge, New York, New Rochelle: Cambridge University Press, 1989), 461–7, and passim; Adele E. Clarke and Joan H. Fujimura, eds., *The Right Tools for the Job: At Work in Twentieth-Century Life Sciences* (Princeton, N.J.: Princeton University Press, 1992), 31–44, and passim; and Albert Van Helden and Thomas L. Hankins, eds., *Instruments* (= *Osiris*, second ser. 9 [1994], on 237–42). On optical instrumentation in particular see Gerard L'E Turner, "The History of Optical Instruments: A Brief Survey of Sources and Modern Studies", *History of Science 8* (1969):53–93; reprinted with supplement in idem, *Essays on the History of the Microscope* (Oxford: Senecio, 1980), chapter 2.

[2] David S. Landes, *The Unbound Prometheus: Technological Change and Industrial Development in Western Europe from 1750 to the Present* (Cambridge: Cambridge University Press, 1972), 231–358, esp. 326–58; and Otto Keck, "The National System for Technical Innovation in Germany", in *National Innovation Systems: A Comparative Analysis*, ed. Richard R. Nelson (New York and Oxford: Oxford University Press, 1993), 115–56, esp. 115–30.

[3] The only previous study of the ultramicroscope is the brief and unsatisfactory account of mostly post-1903 developments by Hans Gause, "The Slit Ultramicroscope after Siedentopf and Zsigmondy – An Historical and Optical Study", *Jena Review 11:6* (1966):327–33.

[4] This is, of course, not to say that there were no German manufacturers of microscopes and other optical instruments, on which see, e.g., Hans Weil and Helmut Baden, "Schieck and the Beginnings of the German Microscope Industry", *Bulletin of the Scientific Instrument Society* No. 18 (1988):9–12. For an overview of the European-wide trade in microscope manufacturing see Gerard L'E Turner, "Making Microscopes: The Nineteenth-Century European Trade", in idem, T*he Great Age of the Microscope: The Collection of the Royal Microscopical Society through 150 Years* (Bristol and New York: Adam Hilger, 1989), 7–17.

[5] Brücke to Helmholtz, 19 Oct 1857, Helmholtz *Nachlass*, Berlin-Brandenburgische Akademie der Wissenschaften.

[6] Unless otherwise noted, the following account of Zeiss relies on Felix Auerbach, *Ernst Abbe. Sein Leben, sein Wirken, seine Persönlichkeit*, 2nd ed. (Leipzig: Akademische Verlagsgesellschaft, 1922), 115–22; Friedrich Schomerus, *Geschichte des Jenaer Zeisswerkes 1846–1946* (Stuttgart: Piscator, 1952), 1–23; and Horst Alexander Willam, *Carl Zeiss 1816–1888* (Munich: F. Bruckmann, 1967). See also Moritz von Rohr (with Max Fischer and August Köhler), "Zur Geschichte der Zeissischen Werkstätte bis zum Tode Ernst Abbes", *Forschungen zur Geschichte der Optik 1* (= Beilagehefte zur *Zeitschrift für Instrumentenkunde*), H. Boegehold, et al., eds. (Berlin: Julius Springer, 1936), 91–201. (A second, enlarged edition of this work by Rohr et al., appeared in *Forschungen zur Geschichte der Optik 2* [= Beilagehefte zur *Zeitschrift für Instrumentenkunde*], H. Boegehold, et al., eds. [Berlin: Julius Springer, 1938], 1–119.) In addition, see also Moritz von Rohr, "Über die Arbeitsgemeinschaft von Carl Zeiss und Ernst Abbe bis zum Ende der siebziger Jahre I", ibid., 160–76; idem, "Über den Ausgang der Arbeitsgemeinschaft von Carl Zeiss und Ernst Abbe bis zum Ende der siebziger Jahre II", ibid., 253–92; idem, "Ernst Abbe als Leiter der Werkstätte bis zu seinem Tode. III", ibid., 295–346; and Hans Gause, "Carl Zeiss: On the 150th Anniversary of his Birthday, 11th September 1966", *Jena Review (Supplement) 11:4* (1966):2–12.

[7] Zeiss quoted in Schomerus, *Geschichte*, 5.

[8] Auerbach, *Abbe*, chapter 2; and Schomerus, *Geschichte*, 24. Unless otherwise noted, the following discussion of Abbe's early life and career draws on Auerbach, *Abbe*, chapter 2. Although quite old and hagiographic in its approach and tone, Auerbach's biography nonetheless remains the best source on Abbe. See also anon., "Ernst Abbe (1840–1905). The Origin of a Great optical Industry", *Nature 145:3664* (20 Jan 1940):89–91; Norbert Günther, *Ernst Abbe: Schöpfer der Zeiss Stiftung* (Stuttgart: Wissenschaftliche Verlagsgesellschaft, 1951); E. Brüche, "Ernst Abbe und sein Werk", *Physikalische Blätter 21* (1965):261–9; Harald Volkmann, *Carl Zeiss und Ernst Abbe. Ihr Leben und ihr Werk* (Munich: VDI Verlag, 1966); Joachim Wittig, *Ernst Abbe. Sein Nachwirken an der Jenaer Universität. Zu seinem 150. Geburtstag am 23. Januar 1990*, Jenaer Reden und Schriften (N.p.: N.p., 1989); and Joachim Wittig, *Ernst Abbe* (Leipzig: BSB B.G. Teubner, 1989).

[9] Auerbach, *Abbe*, 37–8; Max Steinmetz, et al., eds., *Geschichte der Universität Jena 1548/58-1958. Festgabe zum vierhundertjährigen Universitätsjubiläum*, 2 vols. (Jena: VEB Gustav Fischer, 1958), 1:410–11; W. Schütz, "Die Physik an der Universität Jena im Wandel ihrer Zeit", in *Beiträge zur Geschichte der Mathematisch-Naturwissenschaftlichen Fakultät der Friedrich-Schiller-Universität Jena anlässlich der 400-Jahr-Feier* (Jena: VEB Gustav Fischer, 1959), 9–32, esp. 21–4; and Christian

Heermann, "Karl Snell und Hermann Schäffer als Hochschulpädagogen. Zur Geschichte des Experimentalunterrichtes in Physik an der Universität Jena in der zweiten Hälfte des 19. Jahrhunderts", *NTM 2:6* (1965):23–36.
[10] Auerbach, *Abbe*, 45.
[11] Ibid., 46, 48, 56–7.
[12] Ibid., 69–70, 74–82. On the Jena physics institute see also David Cahan, "The Institutional Revolution in German Physics, 1865–1914", *Historical Studies in the Physical Sciences 15:2* (1985):1–65, on 9, 13–14.
[13] Ernst Abbe, *Ernst Abbe. Briefe an seine Jugend- und Studienfreunde Carl Martin und Harald Schütz 1858–1865*, eds. Volker Wahl und Joachim Wittig, et al. (Berlin: Akademie-Verlag, 1986), 267–75, quote on 268–89. Cf. Abbe to Schütz, ibid., p. 296.
[14] Auerbach, *Abbe*, 115; and Schomerus, *Geschichte*, 32.
[15] Auerbach, *Abbe*, 86–7, 147; Schomerus, *Geschichte*, 47–8, citing a letter from Karl Snell to a friend, 30 May 1878; and Schütz, "Physik an der Universität Jena", 22.
[16] Volkmann, "Ernst Abbe and His Work", 1722, implicitly alludes to the division of labor at Zeiss. I thank Gerry Martin for first drawing my attention to this important point.
[17] Abbe's classic, single most important study of the theory of the microscope was his "Beiträge zur Theorie des Mikroskops und der mikroskopischen Wahrnehmung", *Archiv für mikroskopische Anatomie 9* (1873):413–68. This and his other papers (only some of which were previously published) are (re)printed in *Gesammelte Abhandlungen von Ernst Abbe*, 5 vols. (Jena: Gustav Fischer, 1904–40), vol. 1: *Abhandlungen über die Theorie des Mikroskops* (1904). Analyses of Abbe's theory include M.J. Michael, "Numerical Aperture Reconsidered", *Journal of the Royal Microscopical Society 15* (1895):609–23; Auerbach, *Abbe*, 122–46; Royal Microscopical Society, "Discussion on the Abbe Theory", *Journal of the Royal Microscopical Society 49* (1929):123–42, 228–64, which is a set of eleven papers devoted to Abbe's theory of image formation and the resolving power of the microscope; H. Volkmann, "Ernst Abbe and His Work", *Applied Optics 5:11* (1966):1720–31, esp. 1723–4; S. Bradbury, *The Evolution of the Microscope* (Oxford, London, Edinburgh: Pergamon Press, 1967), 240–5; N. Günther, "Abbe, Ernst", in *Dictionary of Scientific Biography 1* (1970), ed., Charles Coulston Gillespie, 16 vols. (New York: Charles Scribners, 1970–80), 6–9; and Kei-ichi Tsuneishi, "On the Abbe Theory (1873)", *Japanese Studies in the History of Science 12* (1973):79–91, which compares and contrasts Abbe's theory with a seemingly similar theory proposed by Hermann von Helmholtz in 1874. Kei-ichi shows, moreover, that until 1900 many scientists (mistakenly) viewed Helmholtz's and Abbe's theories of the microscope as similar and that most favored Helmholtz's. See also Bernd Wilhelmi, *Lichtmikroskopie. Ernst Abbe und sein Einfluss auf moderne Entwicklungen* (Berlin: Akademie Verlag, 1991) (= *Sitzungsberichte der Sächsischen Akademie der Wissenschaften zu Leipzig. Mathematisch-naturwissenschaftliche Klasse 123:2*).
[18] On the basis of Abbe's theory of image formation, Otto Wiener nominated Abbe in 1904 for the Nobel Prize in Physics (Otto Wiener to Nobelkommitté für Physik der Königl. Akademie der Wissenschaften zu Stockholm, 27 Jan 1904, 1904 Fysik.); and Adolf Winkelmann did the same on the basis of Abbe's overall contributions to microscopy, noting, i.a., his theory of image formation as a consequence of diffraction (Adolf Winkelmann to the Nobel-Comité der Physik der Königlichen Akademie der Wissenschaften in Stockholm, 28 Dec 1904, 1905 Fysik), both in Royal Swedish Academy of Sciences, Stockholm (hereafter RSASS).
[19] Schomerus, *Geschichte*, 32–5, 44; and Bradbury, *Evolution*, 225, 227, 256, 273.
[20] Wolfgang Mühlfriedel and Edith Hellmuth, "Die Geschichte der Optischen Werkstätte Carl Zeiss in Jena von 1875 bis 1891", *Zeitschrift für Unternehmungsgeschichte 38* (1993):4–25, on 8–13, 18.
[21] Ibid., 5; Auerbach, *Abbe*, 73, 151; idem, *The Zeiss Works and the Carl-Zeiss Stiftung in Jena: Their Scientific, Technical, and Sociological Development and Importance Popularly Described*, trans. from the 2nd German ed. by Siegfried F. Paul and Frederic J. Cheshire (London: Marshall, Brookes, & Chalkley, 1904), 136; and Schomerus, *Geschichte*, 38–42, 68.
[22] Abbe quoted in Auerbach, *Abbe*, 158; cf. 160.
[23] Ernst Abbe, "Neue Apparate zur Bestimmung des Brechungs- und Zerstreuungsvermögens fester und flüssiger Körper", *Jenaischen Zeitschrift für Naturwissenschaft 8* (1874):96–174, reprinted in his *Gesammelte Abhandlungen 2, Wissenschaftliche Abhandlungen aus verschiedenen Gebieten. Patentschriften. Gedächtnisreden* (Jena: Gustav Fischer, 1906), on 82–163. See also his report to the Prussian Kultusminister, Adalbert Falk, on his visit in September 1876 to the London International Exhibition, "Die optischen Hülfsmittel der Mikroskopie", in A. W. Hofmann, ed., *Bericht über die Wissenschaftliche Apparate auf der Londoner Internationalen Ausstellung in 1876* (Braunschweig: Friedrich Vieweg und Sohn, 1878), 383–420, reprinted in his *Gesammelte Abhandlungen 1*:119–64.

[24] "Die optischen Hülfsmittel der Mikroskopie", 159, and cited in Volkmann, "Abbe", 1725 and Bradbury, *Evolution*, 259–61 (quote on 260).

[25] Unless otherwise noted, the present account of Schott relies on Schomerus, *Geschichte*, 53–67. See also Auerbach, *Abbe*, 160–79; and Hans-Günther Körber, "Schott, Otto Friedrich", *Dictionary of Scientific Biography 12* (1975), ed., Charles Coulston Gillispie, 16 vols. (New York: Charles Scribners, 1970–80), 211–12, which contains an extensive bibliography.

[26] Auerbach, *Abbe*, 160–1.

[27] See Schott's letter to Abbe of 27 May 1879, reprinted in Schomerus, *Geschichte*, 55.

[28] Herbert Kühnert, ed., *Der Briefwechsel zwischen Otto Schott und Ernst Abbe über das optische Glas, 1879–1881* (Jena: Gustav Fischer, 1946). See also Schomerus, *Geschichte*, 55–8. For further documentary material concerning Abbe's and Schott's efforts to establish their glass factory see Kühnert, ed., *Briefe und Dokumente zur Geschichte des VEB Optik Jenaer Glaswerk Schott & Genossen*, 1. Teil, *Die wissenschaftliche Grundlegung (Glastechnisches Laboratorium und Versuchsglashütte) 1882–1884* (Jena: Gustav Fischer, 1953), and Abbe's *Gesammelte Abhandlungen 4*, entitled *Die Entstehung des Glaswerks von Schott & Gen nach gleichzeitigen Schriftstücken aus amtlichem und persönlichem Besitz zwischen dem März 1882 und dem Januar 1885* (Jena: Gustav Fischer, 1928).

[29] Quoted in Auerbach, *Abbe*, 164–6.

[30] Quoted in ibid., 166–7.

[31] Abbe to Schott, 21 Mar 1881, in H. Kühnert, ed., *Briefwechsel zwischen Schott und Abbe*, 50, quoted in Stuart Michael Feffer, "Microscopes to Munitions: Ernst Abbe, Carl Zeiss, and the Transformation of Technical Optics, 1850–1914", Ph.D. diss., University of California Berkeley (1994), 207.

[32] Auerbach, *Abbe*, 167–9; and Herbert Kühnert, "Ein unbekannter Brief von Ernst Abbe an Hermann v. Helmholtz. Ein Beitrag zur Vorgeschichte des Jenaer Glaswerks Schott und Genossen", *Sprechsaal für Keramik, Glas, Email 94:9* (1961):213–16 and *94:12* (1961):322–3, on 215.

[33] Auerbach, *Abbe*, 169–70; and Schomerus, *Geschichte*, 59–60.

[34] David Cahan, *An Institute for an Empire: The Physikalisch-Technische Reichsanstalt 1871–1918* (Cambridge, New York, New Rochelle: Cambridge University Press, 1989), chapter 2.

[35] Abbe to Helmholtz, 15 May 1883, Helmholtz *Nachlass*, Berlin-Brandenburgische Akademie der Wissenschaften, cited in Kühnert, "Ein unbekannter Brief", 214.

[36] Kühnert, "Ein unbekannter Brief", 216.

[37] Abbe to Nikolaus Kleinenberg, 8 June 1883, quoted in ibid., 323.

[38] Auerbach, *Abbe*, 171–3 (quote on 171–2); and Schomerus, *Geschichte*, 61–3 (quote on 61).

[39] Abbe, "Ueber Neue Mikroskope", *Sitzungsberichte der Jenaischen Gesellschaft für Medizin und Naturwissenschaft* (1886), 107–28, reprinted in his *Gesammelte Abhandlungen 1*:450–72. Abbe's article soon appeared in English translation as: "On Improvements of the Microscope with the Aid of New Kinds of Optical Glass", *Journal of the Royal Microscopical Society 6:2* (1886):20–34. For a discussion of the Werke's development and early marketing of the apochromatic objective see Feffer, "Microscopes to Munitions", 244–54.

[40] Schomerus, *Geschichte*, 45; Bradbury, *Evolution*, 263, 266–7; and Feffer, "Microscopes to Munitions", 185–91.

[41] [Ernst Abbe and Otto Schott], "Productionsverzeichniss des glastechnischen Laboratoriums von Schott und Genossen in Jena", July 1886, reprinted in Abbe's *Gesammelte Abhandlungen 2*: 194–205; Auerbach, *Abbe*, 174–5; and Schomerus, *Geschichte*, 64. The entire chronology of optical glass developments at Zeiss and Schott from 1873 to 1886 is most easily followed in anon., "Zeittafel zum Glaswerk", in Abbe's *Gesammelte Abhandlungen 4*:140–2.

[42] H. Hovestadt, *Jena Glass and Its Scientific and Industrial Applications*, trans. J.D. Everett and Alice Everett (London and New York: Macmillan, 1902), 26–31, 387–93, presents supplementary product lists of glass which in 1888 ran to sixty-eight in number, in 1892 to seventy-six, and in 1902 to sixty-eight.

[43] Volkmann, "Abbe", 1728; and Bradbury, *Evolution*, 263–70. G. L'E. Turner, "The Microscope as a Technical Frontier in Science", in *Historical Aspects of Microscopy*, S. Bradbury and G. L'E. Turner, eds. (Cambridge: W. Heffer, 1967), 175–99, on 188–9, reprinted in Turner's *Essays*, chapter 9, shows graphically the increase in numerical aperture for microscopes during the nineteenth century.

[44] Auerbach, *The Zeiss Works*, 31, gives a slightly different developmental history, dividing it into phases lasting from 1846 to 1872 (i.e., until Abbe's development of his theory of image formation); 1872 to 1889 (a transition period during which homogeneous and apochromatic microscopes, i.a.,

were developed and in which microscope production went from individual artisanal production to that of an organized, large-scale industrial system); and post-1889 (a period of maturity).

[45] Schomerus, *Geschichte*, 152; and Auerbach, *The Zeiss Works*, 136, 138–9. Schomerus and Auerbach give slightly different figures for the number of employees.

[46] Mari E.W. Williams, *The Precision Makers: A History of the Instruments Industry in Britain and France, 1870–1939* (London and New York: Routledge, 1994), 4, 34, 36, 40–2.

[47] Auerbach, *The Zeiss Works*, 76.

[48] Mühlfriedel and Hellmuth, "Die Geschichte der Optischen Werkstätte", 14–15.

[49] Auerbach, *The Zeiss Works*, 88–9; idem, *Abbe*, 257–8; and Schomerus, *Geschichte*, 68–9, 105–8.

[50] Auerbach, *Abbe*, 238–76; and Schomerus, *Geschichte*, 100–3.

[51] Ernst Abbe, "Die Denkschrift vom 4. Dez. 1887", in idem, *Gesammelte Abhandlungen 5*, ed., Friedrich Schomerus, *Werden und Wesen der Carl Zeiss-Stiftung an der Hand von Briefen und Dokumenten aus der Gründungszeit (1886–1896)* (Jena: Gustav Fischer, 1940), 35–78, on 40–1.

[52] Schomerus, *Geschichte*, 104–5.

[53] In addition to the discussion in Auerbach, *Abbe*, 238–76, and in Schomerus, *Geschichte*, 100–26, see also, i.a., Abbe's *Gesammelte Abhandlungen 5, Werden und Wesen der Carl-Zeiss-Stiftung an der Hand von Briefen und Dokumenten aus der Gründungszeit (1886–1896) dargestellt von Fr. Schomerus* (Jena: Gustav Fischer, 1940), 2nd ed. (Stuttgart: Gustav Fischer, 1955); and Ernst Wuttig, "Die Carl Zeiss-Stiftung in Jena und ihre Bedeutung für die Forschung", in Ludolph Brauer, et al., eds., *Forschungsinstitute. Ihre Geschichte, Organisation und Ziele*, 2 vols. (Hamburg: Paul Hartung, 1930), *1*:441–9.

[54] Schomerus, *Geschichte*, 114–16.

[55] Auerbach, *Abbe*, 324; and idem, *The Zeiss Works*, 144.

[56] Auerbach, *Abbe*, 324–7. For example, the Stiftung supported the physics institute, giving it an entirely new building in 1902; a new chair in technical physics and mathematics; and, as mentioned above, a new Institute for Technical Chemistry. It also provided support for new institute buildings or major expansions for minerology, hygiene, pathological anatomy, botany, zoology, anatomy, pharmacology, chemistry, as well as the university's library and surgical clinic. (See anon., "Hauptsächlichste Leistungen der Carl Zeiss-Stiftung für die Universität Jena", in Abbe's *Gesammelte Abhandlungen 5*:278–9; Steinmetz, et al., eds., *Geschichte der Universität Jena*, 478–9, 482–3, 502–3; Auerbach, *Abbe*, 326; and idem, *The Zeiss Works*, 126–7.)

[57] Biographical information about Zsigmondy is drawn from Alfred Lottermoser, "Zsigmondy, Richard", in *Deutsches Biographisches Jahrbuch 11* (1929):335–8; and "Chemistry 1925. Richard Adolf Zsigmondy", in *Nobel Lectures including Presentation Speeches and Laureates' Biographies. Chemistry. 1922–1941* (Amsterdam, London, New York: Elsevier, 1966), 39–59, which includes Zsigmondy's Nobel Prize address: "Properties of Colloids". Nobel Lecture, December 11, 1926".

[58] On Kundt and his style of physics see David Cahan, "From Dust Figures to the Kinetic Theory of Gases: August Kundt and the Changing Nature of Experimental Physics in the 1860s and 1870s", *Annals of Science 47* (1990):151–72.

[59] Zsigmondy quoted in Lottermoser, "Zsigmondy", 335–6.

[60] Zsigmondy to Svante Arrhenius, 12 Dec 1894, in Alois Kernbauer, ed., *Svante Arrhenius' Beziehungen zu Österreichischen Gelehrten: Briefe aus Österreich an Svante Arrhenius (1891–1926)* (Graz: Akademische Druck- u. Verlagsanstalt, 1988) (= Publikationen aus dem Archiv der Universität Graz, Band 21), 98.

[61] Richard Zsigmondy, *Zur Erkenntnis der Kolloide. Über irreversible Hydrosole und Ultramikroskopie* (Jena: Gustav Fischer, 1905), translated into English as *Colloids and the Ultramicroscope. A Manual of Colloid Chemistry and Ultramicroscopy*, trans. Jerome Alexander (New York: John Wiley; London: Chapman & Hall, 1909), quotes (from the English edition) on vii and v–vi, respectively.

[62] *The Chemistry of Colloids*, Part I, Richard Zsigmondy, *Chemistry of Colloids*, trans. Ellwood B. Spear; Part II, Ellwood B. Spear (with a chapter by John Foote Norton), *Industrial Colloidal Chemistry* (New York: John Wiley; London: Chapman & Hall, 1917), 5–6.

[63] In chapter 3 of *Colloids and the Ultramicroscope*, Zsigmondy cited extensive sections of Graham's classic papers of 1861 and 1862 on colloids.

[64] Zsigmondy, *Colloids and the Ultramicroscope*, 11, 17.

[65] Idem, "Properties of Colloids", 47–8.

[66] Ibid., 46–8 (quotes on 47–8); and idem, *Colloids and the Ultramicroscope*, 65–89.

[67] Zsigmondy, "Properties of Colloids", 48–9. The third question concerned the low level of osmotic energy emitted in gold solutions; it was later addressed experimentally by Svedberg and Perrin.

[68] Zsigmondy, *Colloids and the Ultramicroscope*, 97–8.

[69] Ibid., 98–100, quote on 98. Turner, "The Microscope as a Technical Frontier in Science", 189–90, shows that the highest resolution achieved prior to 1900 was just under 0.2 μm.

[70] Zsigmondy, Colloids and the Ultramicroscope, 100.

[71] Hans-Günther Körber, "Siedentopf, Henry Friedrich Wilhelm", in Dictionary of Scientific Biography 12 (1975), ed., Charles Coulston Gillispie, 16 vols. (New York: Charles Scribners, 1970–80), 12:422–3; and Auerbach, The Zeiss Works, 142–3.

[72] Schomerus, Geschichte, 154; Auerbach, The Zeiss Works, 142–3; and Feffer, "Microscopes to Munitions", 264–6, 271, 285–90, 311.

[73] M. v. Rohr, "Erinnerungen an Ernst Abbe und den Optikerkreis um ihn", Die Naturwissenschaften 6:22 (1918):317–22; ; 6:23 (1918):337–42, on 339–40.

[74] Auerbach, Abbe, 196; and Schomerus, Geschichte, 161.

[75] Zsigmondy, Colloids and the Ultramicroscope, 102, n. 1. Cf. M. v. Rohr, "Erinnerungen an Ernst Abbe und den Optikerkreis um ihn", 341.

[76] H. Siedentopf and R. Zsigmondy, "Über Sichtbarmachung und Grössenbestimmung ultramikroskopischer Teilchen, mit besonderer Anwendung auf Goldrubingläser", Annalen der Physik 10 (1903), 1–39, on 1.

[77] Ibid., 39. The paper was finished in October 1902, and received by the Annalen's editor on 13 October 1902. Cf. H. Siedentopf, "On the Rendering Visible of Ultra-Microscopic Particles and of Ultra-Microscopic Bacteria", Journal of the Royal Microscopical Society (October 1903):573–8, who concluded his paper (578) as follows: "In conclusion I must point out that these investigations have been materially assisted by the liberal manner in which all the necessary means were placed at our disposal by the firm Carl Zeiss of Jena".

[78] Zsigmondy, Colloids and the Ultramicroscope, 102.

[79] Zsigmondy to Arrhenius, 1 Mar 1909, in Kernbauer, ed., Arrhenius' Beziehungen, 221. Yet Zsigmondy did not fail to credit Siedentopf for his original contribution in developing the instrument into a working device (above all, for introducing the slit), in developing the theory of rendering the particles visible, and in calculating probable visibility limits for the particles; cf. Zsigmondy, Colloids and the Ultramicroscope, 100–1.

[80] In their first, and classic, paper describing the ultramiscrope and its use, "Über Sichtbarmachung", 2, Siedentopf and Zsigmondy explicitly noted the division of labor.

[81] Ibid., 1–2 (quote on 2).

[82] Ibid., 2–3.

[83] Ibid., 4–6 (quote on 4–5).

[84] Ibid., 6–8.

[85] For details on this see ibid., 2–16; Siedentopf, "On the Rendering Visible", 574–7; and Zsigmondy, Colloids and the Ultramicroscope, 103–8, 111–23.

[86] Siedentopf and Zsigmondy, "Über Sichtbarmachung", 8–10 (quote on 10).

[87] Ibid., 11–12, 14–15.

[88] Ibid., 16–29 (quote on 17). One potentially important source of error was that occasioned by the straightforward counting of the number of particles within a given area. Zsigmondy argued against committing such an error by introducing an old result that Abbe had arrived at (in 1878) for the probable relative deviation in counting blood corpuscles. (Ibid., 23–4.)

[89] Along with the Zeiss Werke, another industrial source that played a (minor) role in the first use of the ultramicroscope was the glass factory of J.L. Schrieber, in Zombkowice, Russia, which produced the gold ruby glasses. (Ibid., 30; and Zsigmondy, Colloids and the Ultramicroscope, 101.)

[90] Siedentopf and Zsigmondy, "Über Sichtbarmachung", 31–2, 35, 37–8.

[91] Zsigmondy, Colloids and the Ultramicroscope, 160–4.

[92] Ibid., 134–40, quotes on 134–5.

[93] Among the several accounts of the history of Brownian motion during the first decade of the twentieth century, see especially G.L. de Haas-Lorentz, Die Brownsche Bewegung und einige verwandte Erscheinungen (Braunschweig: Friedr. Vieweg, 1913); Stephen G. Brush, "A History of Random Processes. I. Brownian Movement from Brown to Perrin", Archive for History of Exact Sciences 5 (1968):1–36; Mary Jo Nye, Molecular Reality: A Perspective on the Scientific Work of Jean Perrin (London: Macdonald; New York: American Elsevier, 1972); Abraham Pais, 'Subtle is the Lord...': The Science and the Life of Albert Einstein (Oxford, New York, and Toronto: Oxford University Press, 1982), 79–107; and John Stachel, et al., "Einstein on Brownian Motion", in Stachel, et al., eds., The Collected Papers of Albert Einstein, vol. 2, The Swiss Years: Writings, 1900–1909 (Princeton: Princeton University Press, 1989), 206–22.

[94] Albert Einstein, "Über die von der molekularkinetischen Theorie der Wärme geforderte Bewegung von in ruhenden Flüssigkeiten suspendierten Teilchen", Annalen der Physik 17 (1905):549–60; idem, "Zur Theorie der Brownschen Bewegung", Annalen der Physik 19

(1906):371–81; idem, "Theoretische Bemerkungen über die Brownsche Bewegung", *Zeitschrift für Elektrochemie und angewandte physikalische Chemie 13* (1907):41–2; idem, "Elementare Theorie der Brownschen Bewegung", *Zeitschrift für Elektrochemie und angewandte physikalische Chemie 14* (1908):235–9; and Marian von Smoluchowski, "Zur kinetischen Theorie der Brownschen Molekularbewegung und der Suspensionen", *Annalen der Physik 21* (1906):756–80. In 1908, Paul Langevin gave a new derivation of Einstein's and Smoluchowski's equivalent results, "Sur la théorie du mouvement brownien", *Comptes rendus 146* (1908):530–3.

[95] Einstein, "Zur Theorie der Brownschen Bewegung", 334; and Stachel, et al., "Einstein on Brownian Motion", 211.

[96] Paris: Masson & Cie, 1906.

[97] Cotton and Mouton, *Les Ultramicroscopes*, 38–57.

[98] Per Stenius, "Svedberg's Early Studies in Colloid Chemistry", in Bengt Rånby, ed., *Physical Chemistry of Colloids and Macromolecules* (Oxford, London, and Edinburgh: Blackwell Scientific, 1987), 17–21, on 18–19; and Anders Lundgren, "The Ideological Use of Instrumentation: The Svedberg, Atoms, and the Structure of Matter", in *Center on the Periphery: Historical Aspects of 20th-Century Swedish Physics*, ed. Svante Lindqvist (Canton, MA: Science History, 1993), 327–45, esp. 327–30, 333. In September 1903, not long after the appearance of Siedentopf and Zsigmondy's article in the *Annalen*, the Zeiss Werke issued its first catalogue for the ultramicroscope. One could buy either the entire apparatus at a cost of 958 marks or one could buy individual components and so construct one's own ultramicroscope. (*Kostenanschlag über den Apparat zur Untersuchung ultramikroskopischer Teilchen nach Siedentopf und Zsigmondy* [1903]. Carl Zeiss, Jena. No. 008. September 1903. I thank Frau Dipl-Ing. Edith Hellmuth, Leiterin des Zentralarchiv, Jenoptik, Jena, for supplying me with copies of this and other catalogues for Zeiss ultramicroscopes. In 1904, Zeiss issued a fourteen-page, English-language catalogue [without prices]: *Description of Optical Appliances to Facilitate Visual Perception of Ultramicroscopic Particles* [N.p. {Jena?}: N.p., 1904].) As one entire unit, the ultramicroscope was probably the most expensive microscope of its day. Yet since colloid and other chemists, as well as biologists and medical scientists, already owned their own microscopes, some doubtless chose only to buy select individual components in order to create their own ultramicroscope. As early as circa 1905, Auerbach claimed that "[a]lthough these 'ultramicroscopes' have been placed on the market for scarcely a year, biologists and medical men have already achieved great results by their aid, to which fact the very large demand that has arisen for these instruments is undoubtedly to be attributed". (Auerbach, *The Zeiss Works*, 44.)

[99] The Svedberg, "Über die Eigenbewegung der Teilchen in kolloidalen Lösungen", *Zeitschrift für Elektrochemie und angewandte physikalische Chemie 12* (1906):853–60.

[100] The Svedberg, "Über die Eigenbewegung der Teilchen in kolloidalen Lösungen. Zweite Mitteilung", *Zeitschrift für Elektrochemie und angewandte physikalische Chemie 12* (1906):909–10.

[101] Einstein, "Theoretische Bemerkungen über die Brownschen Bewegung", quickly pointed out this and other misunderstandings by Svedberg. Later, Jean Perrin also criticized Svedberg's several misunderstandings and the shortcomings of his work; see Perrin's "Mouvement brownien et réalité moléculaire", *Annales de Chimie et de Physique 18* (1909):1–114. For a full critique of Svedberg's misconceptions concerning Einstein's molecular kinetic theory and his confusions deriving therefrom, see Milton Kerker, "The Svedberg and Molecular Reality", *Isis 67:237* (1976):190–216. See also Nye, *Molecular Reality*, 97, 118, 122–5, 127; Stenius, "Svedberg's Early Studies", 17–18, 20; and Stachel, et al., "Einstein on Brownian Motion", 219–20.

[102] For references to Perrin's numerous original papers and for analysis of his work see Nye, *Molecular Reality*, 84–5, 89, 97–142, and passim. See also Haas-Lorentz, *Die Brownsche Bewegung*, 31–44; and Brush, "History of Random Processes", 30–5.

[103] W[ilhelm]. O[stwald]., review of Zsigmondy's *Zur Erkenntnis der Kolloide. Über irreversible Hydrosole und Ultramikroskopie*, in *Zeitschrift für physikalische Chemie, Stöichiometrie, und Verwandtschaftslehre 57* (1907):383.

[104] Wilhelm Ostwald, "Bücherschau", *Zeitschrift für physikalische Chemie, Stöichiometrie, und Verwandtschaftslehre 64* (1908):509; and idem, "Vorbericht", to his *Grundriss der physikalischen Chemie* (Grossbothen: 1908), cited in Nye, *Molecular Reality*, 151–2.

[105] D. Vorländer to the Nobel-Komitee der Chemie . . ., 24 Jan 1913, "1913 Kemi", in RSASS.

[106] Georg Wiegner to Kungl. Vetenskapsakademiens Nobelkommitté för Kemi, 2 Jan 1915, "1915 Kemi", in RSASS.

[107] R. Wegscheider to the Nobelkomitee für Chemie, 25 Jan 1918, "1919 Kemi"; in RSASS.

[108] W. Ostwald to the Nobelkomitee für Chemie, 10 Jan 1921, "1921 Kemi"; W. Ostwald to das Nobelkomitee für Chemie, received 4 Dec 1922, "1923 Kemi" (quote); and W. Ostwald to das Nobelkomitee für Chemie, 12 Jan 1924, "1924 Kemi", in RSASS.

[109] Hugo R. Kruyt to Messieurs, "1921 Kemi", in RSASS. He wrote: "Above all, Zsigmondy put colloidal chemistry on a firm basis through the part that he took in the discovery of the ultramicroscope (1903), after having done some remarkable research on colloidal gold (1898 and following)".

[110] Georg Wiegner to the Nobelkomitée für Chemie, 19 Dec 1921, "1922 Kemi", in RSASS.

[111] K. Alb. Vesterberg to Kungl. Vetenskapsakademiens Nobelkommité för Kemi, 29 Jan 1923, "1923 Kemi", in RSASS.

[112] A. Kötz to die Königliche Akademie der Wissenschaften in Stockholm, 20 Jan 1924, "1925 Kemi", in RSASS.

[113] Viktor Rothmund to the Nobelkomitée für Chemie in Stockholm, 19 Dec 1925; The Svedberg to the Kungl. Vetenskapsakademiens Nobelkommitté för Kemi, 30 Jan 1926; and G. Meyer to Nobelkomitee für Chemie, 27 Jan 1926, all in "1926 Kemi", in RSASS.

[114] Nobel Lectures, 44.

[115] Körber, "Siedentopf", 422–3.

[116] Zsigmondy, Colloids and the Ultramicroscope, 101–2.

[117] Ibid., 3–5.

[118] For another study of the relationship between the local culture of a laboratory and its influence on the outside world see Simon Schaffer, "Late Victorian Metrology and Its Instrumentation: A Manufactory of Ohms", in Robert Bud and Susan E. Cozzens, eds., Invisible Connections: Instruments, Institutions, and Science (Bellingham, Wash.: SPIE Optical Engineering Press, 1992), 23–56.

[119] Cf., e.g., Gooding, Pinch, and Schaffer, eds., Uses of Experiment, 10, who point to this conclusion in the work of J.L. Heilbron, Electricity in the 17th and 18th Centuries: A Study of Early Modern Physics (Berkeley and London: University of California Press, 1979); Ian Hacking, Representing and Intervening: Introductory Topics in the Philosophy of Natural Science (Cambridge, London, New York: Cambridge University Press, 1983); and D.J. de S. Price, "Of Sealing Wax and String", Natural History 93 (1984):48–56. Cf., too, Peter Galison, How Experiments End (Chicago and London: University of Chicago Press, 1987).

[120] Cf. Gooding, Pinch, and Schaffer, eds., Uses of Experiment, 2–5.

[121] Speaking of the metallurgical laboratory at the Werke, Auerbach wrote circa 1905: "New constructions, and special orders for apparatus for the purposes of research and demonstration, of which a large number are received annually from savants, are, of course, treated specially. Although these special orders, in many cases, do not even pay the cost of execution, they are, notwithstanding, invariably accepted for the sake of any new ideas which they may suggest, if they come legitimately within the province of the undertaking [i.e., the Werke]". (Auerbach, The Zeiss Works, 80.)

[122] Nathan Rosenberg, "Scientific Instrumentation and University Research", Research Policy 21 (1992):381–90, esp. 381–2.

[123] See above, 87–8 and 103–4.

[124] Cf. in this regard the general argument of Nathan Rosenberg, "Basic Research in Industrial Context", in Le Università e le Scienze: Prospettive Storiche e Attuali, ed. Giuliano Pancaldi (Bologna: Aldo Martello, 1993), 233–40.

[125] George Wise, "Ionists in Industry: Physical Chemistry at General Electric, 1900–1915", Isis 74 (1983):7–21, on 7–8. Wise cites the work of Charles Rosenberg, "Science and Social Values in Nineteenth-Century America", in idem, No Other Gods: On Science and American Social Thought (Baltimore and London: The Johns Hopkins University Press, 1976), 135–52; Stuart W. Leslie, "Charles F. Kettering and the Copper-Cooled Engine", Technology and Culture 20 (1979):752–78; Leonard Reich, "Industrial Research and the Pursuit of Corporate Security: The Early Years of Bell Labs", Business History Review 54 (1980):503–29; Lillian Hoddeson, "The Emergence of Basic Research in the Bell Telephone System", Technology and Culture 22 (1981):512–44; and John W. Servos, "The Industrial Relations of Science: Chemical Engineering at MIT, 1900–1939", Isis 71 (1980):531–49.

[126] On the complexity of the relations of science and technology in Imperial Germany see Cahan, An Institute for an Empire.

PRECISION, TOLERANCE, AND CONSENSUS: LOCAL CULTURES IN GERMAN AND BRITISH RESISTANCE STANDARDS

INTRODUCTION

Standards, Witold Kula tells us, have social meaning.[1] Bound to dimensions found in everyday life and labor, potently symbolizing the centralizing tendencies of authorities large and small, and setting guidelines for just practice in commerce and trade, standards are strategic loci on which converge several different dimensions of human behavior. But for Kula, that convergence is rich to the degree it occurs in the deeper past of the early modern period when anthropomorphic measures confined to local cultures dominated. In his view, matters changed following the introduction of the metric system at the end of the eighteenth century. Increased precision and accuracy in measurement expanded the geographic range of standards and diluted to the point of eliminating the subjective connotations and associations of standards found in more local cultures. Hence, for Kula, the metric system helped to destroy feudalism and instill modernizing tendencies. More precise standards became, in Kula's opinion, one foundation for large-scale political unity, a conclusion shared by other historians.[2]

There is something specious about Kula's argument. It accepts all too easily Max Weber's pessimistic view that rational procedures demystify the world, stripping it of meaningful values. Precision in measurement did not eradicate the social meaning of standards; it merely shifted that meaning in the modern period. Standards remained tied to social conventions in scientific practice, especially to highly formalized procedures for measuring. More importantly, standards remained legal entities, and as such, they were still constructed in a way that conformed to customary legal practices and interpretations. Simply put, an unbalanced scale remained unfair and illegal. Or, more strikingly, the plethora of new or revised standards that appeared in science and in daily life in the nineteenth century incorporated past legal traditions. Social meaning did not then simply disappear from standards in the modern period.

The persistence of social meaning is in fact found in what Kula considers most decisive in separating the early modern and modern periods: precision measurement. Recent scholarship has amply demonstrated that in and of themselves, precision measures, like other forms of quantification, do not necessarily prevail or command authority.[3] Meaning is actively assigned to them from among the traditions of local cultures. For most of the nineteenth century, attributes of precision in one culture were not necessarily those in another. Well before concerns over new kinds of standards arose, German practitioners used probability theory to understand the meaning of refined, and later precise, measurements. Precision and error were inseparable, and hence

117

J.Z. Buchwald (ed.), Scientific Credibility and Technical Standards in 19th and early 20th Century Germany and Britain, 117–156.
© *1996 Kluwer Academic Publishers. Printed in Great Britain.*

standards definitions were based on both.[4] Alone, precision measurement does not a standard make, a point that historians have sometimes overlooked. Standards construction entails some compromises in measurement – no standard is absolutely perfect – so as to expand the community likely to accept the standard. The bases for and extent of these compromises vary over time and place. Ultimately, though, public credibility is at stake: to be viable, a standard has to have the largest circle of acceptance possible. Yet there can be problems. The persistence of local or even national cultures of exact experiment, for instance, can mitigate chances that a standard will be uniformly accepted simply because standard's quantitative basis was constructed by disagreeable means. The standard's public, in the broadest sense, might then be constrained.

Among the new modern standards of the nineteenth century, none were more problematic or controversial than that of electrical resistance. Crucial to the commercialization of electricity, the electrotechnology industry, and to scientific practice alike, the standard of electrical resistance was the subject of intense debate for over fifty years, especially in Great Britain and Germany. In 1898 the government of the German Reich finally legalized a unit of electrical resistance, the ohm, specified as "the resistance of a mercury column containing 14.4521 grams of mercury and having a length of 106.3 centimeters with a uniform cross-section of one square millimeter."[5] A slightly elongated version of Werner Siemens's practical mercury unit, dating from the late 1850s, had prevailed as the nation's standard. The ohm's earlier theoretical definition, expressed in terms of Wilhelm Weber's system of absolute units of length, mass, and time, was not cited, even though it had been the foundation of the British resistance measure. National differences had been apparent from the start. In Britain standard resistances were metal coils; in Germany, mercury columns. In both countries, physicists and telegraph electricians worked side-by-side, and protocols for measuring resistance were shaped by both the practical needs of telegraphy and the lofty dictates of exact experimental physics; but telegraphy had the edge in Britain, and physics in Germany. United by telegraph cables, both sides were nonetheless divided in opinion for decades over the most appropriate unit, standard, and measure of electrical resistance.

This essay takes up two comparative considerations in assessing the roles of precision, tolerance, and consensus in determinations of the standard of electrical resistance. The first is intercultural, cutting across geographic lines, and concerns German and British protocols used in manufacturing electrical standards. The case of Great Britain, where telegraphy was extremely influential in electrical standards determinations, has already been well explored by several historians.[6] This essay will focus primarily on German determinations of resistance measures, especially between 1860 and 1880 when British–German exchanges were most acrimonious.

The second comparison is intracultural. It addresses the social meaning of different types of standards in Germany, especially the more traditional and older standards of weights and measures. Associated more with trade and commercialization than technology, weights and measures protocols were far more influential in the scientific and legal establishment of German electrical units and standards than were the demands of electrotechnology. The complexion of experimental practices and legal specifications that characterized German weights and measures formed a powerful social tradition that resisted the easy transference of precision measures across cultural lines. Precision measures that help to define standards are thus at first characteristics of local cultures.

EARLY MEASURES OF RESISTANCE IN THE GERMANIES

Before the establishment of the German–Austrian Telegraph Association in 1854, Wilhelm Weber's research dominated the study of resistance measures, especially between 1841 and 1852. In the physical institutes of both Göttingen and Leipzig, Weber developed a style of exact and painstaking research that helped to make electricity a leading sector in the transformation of German physics both theoretically and experimentally, especially through his magisterial *Elektrodynamische Maassbestimmungen* of 1846.[7] The major features of his style are well known.[8] He brought to the study of resistance his appreciation for the fineness (*Feinheit*) and sharpness (*Schärfe*) of data as well as his emphasis upon the quality of an instrument, especially its sensitivity (*Empfindlichkeit*) and reliability (*Zuverlässigkeit*), which he believed were responsible for the quality of the data. The mere fact that an experiment had yielded the desired result was insufficient, in his view, for making claims about the certainty of its conclusions. He argued that the "surety and certainty of the result" was founded in good part on the investigator's thoroughness in reporting the experiment's instruments, how they were used, how the trials were performed and what changes were made, and how the data was taken and reduced.[9] In the Germanies, Weber's opinion was not singular.

Galvanic measurements, exact though they might have been, were sometimes idiosyncratic and so not directly comparable. Moritz Jacobi's 1846 suggestion – that physicists adopt as "a common unit" a uniform length of copper wire – failed for the reason Jacobi knew from the start: impurities in the metal made an "absolute determination" of resistance impossible.[10] Telegraph engineers found his "standard" resistance boxes "too uncertain."[11] Weber, however, held out hope that eventually Jacobi's standard would find general acceptance.[12]

But there was a catch. If the material character of standards varied somewhat from user to user, then some common ground uniting measurements taken with their help had to be constructed. Weber found the answer by thinking in terms of standards in general. Systems of standards had their *Grundmaass*, their foundational standard, in terms of which measurements were made. Jacobi's standard, a copper wire in a box, could have been a *Grundmaass*, Weber believed, if the copper's resistance had remained constant. But it did not. Furthermore, Weber argued in 1851, resistance was not a *Grundmaass* since it, like velocity, could be expressed in terms of other more fundamental quantities. Adopting Gauss's system of absolute measures – where all quantities were expressed in terms of length, mass, and time – Weber expressed resistance in terms of these fundamental units and defined a unit of resistance in terms of Ohm's Law as that "which a closed conductor possesses when a unit measure of electomotive force produces a unit measure of current intensity." To put this definition in practical working terms, Weber measured the resistance of a "very long and thick" copper wire according to two different methods, found the results to be similar, and then reduced the resistance measures of Jacobi and others to his own absolute units (Jacobi's turned out to be 5,980 million absolute units, or rounded off, 6×10^9 absolute units).[13]

Returning to the problem of resistance measure again a year later, Weber elaborated upon his earlier work. His thinking on measures this time paralleled so closely edicts concerning the more customary weights and measures used in daily life and controlled by central agencies of power, that it is difficult to imagine that he had not been influenced by them, either through his own early interest in standards determinations[14] or perhaps through Gauss, whom he had just rejoined at Göttingen in 1849 and who had been responsible for placing Hannover's weights and measures on a more accurate foundation in the 1830s during Weber's years of exile in Leipzig.[15] Like other measures, Weber argued, standards of resistance are founded on: 1) a definition of the type of quantity to be measured; 2) a specific standard; and 3) a method for comparing measurements of the same quantity.[16] Weber had satisfied the first desiderata with his electrodynamic definition of resistance based on Ohm's Law. The second, insufficiently considered by Jacobi, Weber believed was satisfied by expressing resistance in terms of his *Grundmaasse*: his absolute units of length, weight, and time. Even though he did not mention the one-of-a-kind embodiments of weights and measures called *Urmaasse* kept under lock and key by governments, this expression of resistance in terms of absolute units, although having no physical embodiment, functioned in the same way as did a state's *Urmaasse* because it became the foundation for all functional standards. Then in a manner paralleling the construction of copies of the *Urmaasse*, called *Normalmaasse*, which were distributed to regional administrative units and used to calibrate weights and measures (and their apparatus) used in daily transactions, Weber satisfied his third condition by constructing an absolute resistance unit made of copper wire.[17] Since this *Etalon*, as Weber called it, remained such only for as long as its resistance remained constant, it was somewhat ephemeral; but Weber found his remained constant long enough for him to make copies. He deposited three *Etalons* in Leipzig's physical institute, and then compared them to some of Jacobi's resistance boxes. The hierarchical arrangement Weber used in the definition and construction of the standard followed closely that used in legal weights and measures determinations in all major German states.

Weber never wanted to remove Jacobi's resistances from public use.[18] Jacobi, Weber believed, just had not thought through the ways in which his resistance could be used as a standard. The material form of his own absolute resistance units were, however, special constructions, like the state's *Urmaasse*:

they were difficult to construct, required special settings for their manufacture (what Weber called the "favorable circumstances" of his own laboratory with its complicated exact instruments), and were few in number.[19] They could therefore never enter into common use. What Weber found in Jacobi's resistance box was a standard that could become commonplace. One used absolute measures for resistance, then, only to the extent that Jacobi's resistances had to be calibrated in terms of a special absolute resistance standard. His strategy seemed to suggest that he was relegating Jacobi's resistance box to the status of a copy which has to be recalibrated and tested from time to time so as to certify its agreement with the original.

There were three problems with Weber's system. First, like Jacobi's resistance boxes, even Weber's resistance standards became inaccurate over time. Second, he had a foundation for the *Urmaass*, or a foundational standard, for his system only in the sense that there was a *definition* of resistance in absolute units *and a protocol* for calibrating any wire in terms of those units. His foundational standard was an abstract one, much like the meter was. For that reason the Siemens physicist Oskar Frölich could argue that Weber placed resistance determination on "an unshakable foundation" since the foundation itself did not change. Yet it was precisely that foundation that created the third problem. Weber's instruments and the skill one needed to use them aside, his calculations were so complicated and laborious that German technicians considered them "too fussy".[20]

Weber had thus eliminated exactly the group he wanted to incorporate. He also failed to achieve the goal he had set for himself since the early 1830s: creating means for comparing data, this time especially among technicians. He had had a longstanding interest in data comparison. His and Gauss's telegraph, constructed in 1833 by running a wire from Gauss's observatory to Göttingen's Johanniskirche to Weber's physical institute, was to allow for simultaneous and comparable observations. His large-scale earth-magnetic project from the late 1830s was based on the assumption that measurements made at different locations could be made comparable, and Weber proposed more than one means for doing so.[21] One purpose of his *Elektrodynamische Maassbestimmungen* was to provide instructions for the technical application of galvanism. He bemoaned the fact that technicians worked out their own trials and expressed their own results without any apparent common foundation, thereby wasting time and money by repeating almost the same thing over and over again. Weber believed that his work on the absolute measure of resistances gave technicians what they needed: a way "to explain with few words and numbers, the results of the experiments performed and to give definite and precise instructions for future use".[22] Possessing a common language and method, technicians would then find it easier to compare data. But convincing technicians of the usefulness of his system proved difficult.

Not until 1858 did Germany's telegraph journal, the *Zeitschrift des deutsch-österreichischen Telegraphen-Vereins*, mention the usefulness of Weber's absolute units. An application of Weber's units, and high praise for the system that went with it, appeared in a German translation of William Thomson's article on the conductivity of commercial copper.[23] Thomson's opinion appears to have been singular, despite the overwhelming mathematical, theoretical, and exact experimental orientation of the *ZTV*, its editor, and its authors. The editor, Philipp Wilhelm Brix – nephew of the director of Prussia's bureau of weights and measures, Franz Neumann's first doctoral student at Königsberg, and civil servant and then engineer in the Prussian telegraph administration – was known for bringing to technical problems the theoretical and exact experimental methods of physics. When the depletion of Prussia's forests forced the substitution of coal for wood as the main source of heat, it was Brix whom the *Verein zur Beförderung des Gewerbfleisses* asked to report on the heating power of fuels, with special attention to industrial uses. After 1861 Brix taught applied subjects at Berlin's *Telegraphenschule* and *Bauakademie für Telegraphie* and, from 1876 to 1880, supervised the construction of a large network of underground telegraph lines in Germany. He was known to marvel others with the wonders of exactitude in technology. When late in his life, an underground telegraph line failed, he used refined methods to determine exactly where the break occurred and, to prove he was right, travelled himself to the precise location in the village his calculations had led him to. The subsequent dig vindicated him beautifully. A colleague found it "significant for Prussian and German telegraphy that a physicist from the rigorously theoretical school of Franz Neumann" supervised, as chief telegraph engineer, the intensive development of Germany's telegraph lines.[24]

Brix, ever one to offer editorial comments on the articles he published, brought to the journal several serving didactic purposes. Essays on Ohm's law with applications to telegraphy, on the laws of electromagnetism, and on instruments such as the galvanometer, the *Tangentenbussole*, and the rheostat, ran side-by-side with those on simultaneous transmission, line insulation, fault location, and resistance determination. Not all were written by physicists; telegraph engineers also contributed their share to the theoretical and practical understanding of the tools of their trade. Brix tended, however, to place the greatest weight on what physicists said and did. When a telegraph inspector from Hannover wrote on the insulation of above-ground lines, Brix remarked that the Ministry of Trade had recently set up a special commission, including "two of [Prussia's] most important physicists", to study the matter, so he would withhold any judgment until the commission's work had been completed.[25] The interaction between theory and practice that he advocated was not new in German telegraph engineering: Siemens claimed that he too had promoted knowledge of electrical laws among his "practical electricians" since 1847.[26]

To the extent that the journal offers a view of telegraph practice, it appears that before 1860 resistance determination varied in substance and rigor. In his 1855 article on applications of Ohm's Law in telegraphy, Wilhelm Beetz used as a unit of resistance that "of an iron wire one [Prussian] foot long and one *Linie* cross-section", a choice he knew was "arbitrary" because the characteristics of the wire changed. His calculated resistances were at best comparative ratios, but he considered them good enough in telegraphy. In addition to the galvanometers telegraphers used, Beetz did recommend, however, adding a rheostat.[27] It was the rheostat, as the Passau telegraph electrician Christoph Bergeat demonstrated in 1857, that refined the quantitative element in current measurement and enabled one to speak of electrical measurements in terms of their accuracy. Although acknowledging the practical advantages of the rheostat – one could understand more clearly how changes in resistance affected current strength as well as duplicate resistance units then in use – Bergeat seemed most interested in how it could be used to achieve "all desirable accuracy". He went so far as to point out that for "great accuracy", one would have to know the errors of observation, a "time-consuming" but nonetheless essential task.[28] More accurate were the later rheochords developed independently by Johann Poggendorff, Franz Neumann, and Wilhelm Eisenlohr; the latter's acted like a system of variable weights.[29]

But most of these instruments – and even ideas on their use – were developed by individuals more familiar with the smaller wires found in a laboratory. What kind of accuracy could one really expect out in the field, where telegraph lines were orders of magnitude larger? For German electricians and physicists, the problem of accuracy became crucial in locating faults. Carl Siemens, working from the St. Petersburg office of Siemens & Halske, found in 1855 that using a galvanoscope to locate faults was grossly inadequate, time-consuming, and costly in Russia where telegraph stations were widely separated. In Russia the network over which fault location was conducted had expanded markedly between 1853 and 1855 when the firm, which had originally constructed Russia's telegraph lines in 1851, laid a submarine cable from St. Petersburg to Kronstadt, and then linked the Prussian border to St. Petersburg, Kiev, Moscow, Odessa, and other cities. "A still more accurate determination of faults from the telegraph station was necessary", Siemens stressed with respect to this growth. So in 1856 he deployed a new instrument developed at Siemens, a differential galvanometer; with this he felt he could quantify the accuracy of the fault location. But his accuracy was nowhere near what Bergeat seemed to imply was needed. "In all of the cases", Siemens reported, "the position of the fault was determined so accurately that in the most unfavorable case a deviation of at most 10% of the total line length was produced." Admittedly, that could amount to miles of wire, but if one really wanted "greater certainty", then, according to Siemens, one should just plan to conduct the trial at both stations well enough for both results to "completely agree".[30]

The quantification of resistance took another turn altogether in the late 1850s when, in the Berlin *Experimentirzimmer* of Siemens & Halske, Ernst Esselbach applied the first form of what came to be known as the Siemens unit of resistance: the resistance of a mercury column one square millimeter cross-section and one millimeter long. In practical telegraphic terms, that meant that the resistance of a line made of a Prussian mile of iron wire of 2-$\frac{1}{8}$ *Linien* diameter equalled 64,000 of these units.[31] Esselbach, a former student of Weber and Helmholtz who had joined the telegraph division of Siemens in 1858 after receiving his doctorate in physics under Gustav Karsten at Kiel, tested in the laboratory and calibrated to the Siemens unit samples of several submarine cables, including the Red Sea and transatlantic cables.

"THE INDEPENDENCE OF THE ENGLISH SCHOOL OF ELECTRICIANS OF THAT OF GERMANY"

Werner Siemens's unit dominated German discussions on resistance measures for four decades following its announcement in 1860.[32] Esselbach's unit proved to be impractical for large resistances, so the unit that Siemens finally announced was a mercury column one meter long, 1 mm^2 cross-section, at 0°C. Like Weber before him, Siemens's argument in support of his unit drew upon the guidelines of both an exact experimental physics and weights and measures specifications; but unlike Weber, Siemens made his unit suitable for practical use. The lengthening of the column; the choice of mercury, whose conductivity was relatively constant and less dependent upon temperature than other metals; the "convenient" numbers that the dimensions of his apparatus were intended to produce; the "ease" of the unit's manufacture; all these, Siemens believed, made his unit suitable for "technical physics".

Yet his prescriptions on the construction of the unit were aimed at the physicists who would be responsible for producing it and its copies. So his instructions emphasized the quality of the data (they had to be *Schärfe*, or characterized by a sharp edge, precise and exact); the computation of errors; and the accuracy of the result. Despite his criticism that the construction of Weber's absolute unit of resistance required "very perfect instruments", "a special location", and "great manual dexterity", the manufacture of Siemens's unit could not be accomplished with less means – and could even require more. His argument that the resistance unit should be expressed in geometrical terms evoked parallels to the legal construction of weight standards; but it was also a way to introduce idealized mathematical constructions into his calculations that would eliminate the need for tedious and error-ridden measurements, such as determining the exact diameter of the tube. Hoping at first to work with manufactured glass tubes, he found them too irregular to determine the resistance standard with "sufficient accuracy". Improving the construction of the tube solved only part of the problem; for Siemens found that idealizing the tube as a truncated cone whose volume was determined by weighing the mercury increased the accuracy of his data. He used the finest instruments available: an "accurate chemical balance"; a Wheatstone bridge made "suitable for very accurate resistance measures" by Siemens & Halske; and a *Spiegelgalvanometer* of the type used by Weber. Cognizant of the errors plaguing his method, he either computed them or eliminated them through an artful manipulation of his apparatus and materials. "It cannot be doubted", he boasted, "that the method used is suited for reproducing the resistance standard to any degree of accuracy".[33] That probably overstated his accomplishment; for as he knew, he had not eliminated inaccuracies arising from impurities in the mercury. In the construction of his standard Siemens claimed that he was dealing with a situation where the sensitivity of electrical instruments had already outstripped that of the existing standards used to gauge what those instruments measured. Creating a new resistance standard was a way to align the two. His standard was practical, too; but in the end, technicians were mere recipients of a theory-based, exactly constructed standard intended to complement other electrical apparatus already in common use.

Siemens's thinking on the mercury unit appears to have been guided by how it would eventually fit into a legal system of weights and measures. When he spoke about resistance units and standards, he classified them according to the nomenclature of German weights and measures systems. To check Jacobi's resistance unit, for instance, he assumed that Jacobi had had a *Normalmaass* that he could use; but none was available. When considering a replacement for Jacobi's unit, Siemens thought first of the problem of the *reproducibility* of a standard. Since metals generally changed over time, wire could not be used for an *Urmaass*: not only would the integrity of the *Urmaass* be in doubt, but also *Normalmaasse* based on it would deviate from one another over time, as would copies based on them. Hence for both the *Urmaass* and the *Normalmaasse* Siemens recommended mercury because it changed considerably less than other metals. The *Urmaass* was the original column of mercury; the *Normalmaasse*, more convenient mercury-filled glass spirals of one meter length. Both were difficult to construct and calibrate; neither was intended for common use. Instead, the mercury spirals were the basis for the manufacture of copies, *Etalons*, made of German silver, which Siemens recommended for daily practice.[34] In 1863 Siemens had available for distribution to physicists and telegraph engineers more complex *Etalons*, resistance scales, consisting of a range of his resistance units from 1 to 10,000 "ordered like a system of weights".[35]

Siemens's quasi-legal reasoning and with it, the hierarchical expression of his unit, were lost on the British, when, in the English translation of his article, various types of *Maasse* were rendered without distinction as "standards".[36] The confusion fueled the subsequent metals controversy with the British as well as the well known, decades long exchanges with members of the British Association Committee on Electrical Standards who, in 1862, came out in favor of Weber's absolute unit of resistance. Although each side was guilty of accusing the other of flaws and weaknesses in resistance unit construction that neither side had really fully eliminated, the debates are not without their merit in the history of precision, especially as it relates to standards definitions. There was some common ground. Both sides acknowledged the need for a resistance unit that could be used in technical practice. Some key players – Esselbach, Siemens, and even Gustav Kirchhoff – even recognized that mediation was immediately at hand if all sides accepted the Siemens mercury standard expressed in absolute units.[37] But each side was really talking in a different language. Agreement was not forthcoming on either the parameters of the standard that would later be central to any legal definition, or on the matrix of practices constituting exact measurement. The upshot: each side viewed precision, standards, and the relationship between the two in different ways.

Secure in their belief in the superiority of Weber's theory-based resistance standard, the British argued for the near coalescence between the standard's specifications and precision measurement at the level of both rhetoric and practice. Specifying the "true absolute unit" as "10 millions of metre/second", the British Association Committee argued that any material standard chosen to represent the unit had to be "perfectly definite", possess "absolute permanency", and "should be reproducible with exactitude". Although a very slight difference between the true and actual measures *might* exist, once the standard was set, the Committee believed, it "shall under no circumstances be altered in substance or definition".[38] The metals controversy, present ever since resistance standards were first proposed but intensified after Siemens's proposal, was sustained on the British side by this intense commitment to the perfectibility of a standard. It persisted undiminished even in light of evidence that no metal could be made absolutely pure and its electrical properties remain absolutely constant.[39] Although the British seemed to relax a bit when they allowed for the possibility of a "provisional standard" made of German silver wire if results could be made to agree to within 1%, they remained steadfast in their grasp for accuracy. In the end, their commitment to perfection in the face of unavoidable material imperfection led them to a solution unheard of in the legal establishment of standards hitherto. Their resistance standard took the form of *several* "equal standards, constructed of such metals as promise the greatest constancy", whose permanence of equality would be "rigorously tested". In 1865 the British constructed ten standards – two each of platinum–silver, gold–silver, platinum–iridium, platinum, and mercury – and deposited them at the Kew. "With so many coils for reference, made of such different metals", they argued, "it appears quite improbable that the unit now proposed should be lost". From among these, the "best known construction" would be issued to the public.[40]

Siemens found not "the scientific harmony of systems of measurement" decisive in the choice of a resistance unit, but rather its "practical advantages" and its amenability to a "simple geometrical" expression.[41] Consistently he retained in his standards definition the hierarchical arrangement that characterized the legal expression of standards in the Germanies. For Siemens the production of the original standard had to be secure and protected; hence the original was *designed* to be rare and difficult to reproduce. Reproduction of the original standard was itself a hierarchical process. The *Urmaass* was not reproduced, but replaced with difficulty; its protocols were complex. The *Normalmaasse*, used for *Etalons*, had to be more easily reproducible since their copies entered into common use.

Siemens made his views clear in the metals controversy with Augustus Matthiessen, the German-educated chemist who was a member of the B.A. Committee on Electrical Standards. In objecting to mercury Matthiessen failed to recognize the role Siemens had assigned to the unit-defining mercury standards and erroneously assumed that the mercury columns were "to remain continously in use as resistance *Etalons*". "That is absolutely not the case", Siemens replied, reminding Matthiessen of the public, practical function of his German silver standards.[42] Matthiessen failed, too, to appreciate that the special laboratory conditions required for the manufacture of the mercury standards were necessary only for the production of the original standards.[43] Here and elsewhere the British demonstrated that they distinguished less clearly the levels of production that Germans found so important in defining standards. Nor did they originally differentiate levels of reproduction. Reproduction, the committee thought, was "comparatively easy of accomplishment" a process needed so that "if the original be lost or destroyed it may be replaced" *and* so that *"men unable to obtain copies of the true standard may approximately produce standards of their own".*[44] Essentially, then, the British were willing at first to remove original standards construction from the kind of stricter hierarchical control governing the production of *Urmaasse* in the Germanies, which was in state hands.[45]

For Siemens a resistance standard had to be "manufactured as accurately as the instruments which it [was] supposed to serve". It had to be accurate enough, in fact, to "satisfy the needs of technology".[46] Siemens thus viewed the whole of electrical measuring as a system whose parts came into alignment with one another when each level of accuracy was near equal. In a similar vein, Fleeming Jenkin, like Wilhelm Weber before him, believed that precision in measurement would result in linguistic precision: "the vague terms good and bad conductor or insulator", he explained, "will be replaced, in all writings aiming at scientific accuracy, by those exact measurements" which could be made with an accurate resistance standard.[47] Statements such as these testify to the existence of a culture of precision. Cultures of precision took shape simultaneously with, and sometimes as a result of, weights and measures reform in the nineteenth century.[48] A culture of precision is a shared set of meanings, behavior, and guidelines for decision-making that characterize or accompany precision measurement.[49] Judgments concerning the quality and significance of precision measures are made in the context of these cultures, using the tools available in it.

Both sides in this exchange boasted of how accurate their measurements were. By 1863–64 the British claimed to have come to within 0.08% of the "true value" of the absolute unit; at the same time, Siemens was claiming a minimal accuracy of 0.05% in his mercury unit.[50] But what did these numbers really mean, and why? How were they achieved? Can one simply compare them to one another and conclude that Siemens was more accurate? One aspect of precision measurement that distinguished the British and the Germans in their respective resistance standards determination was the labor that went into it. Historians have already noted that in Britain setting the absolute resistance standard was a strongly cooperative effort.[51] From the start the British were committed to creating "common knowledge" in resistance standards, hoping that they could be reproduced "with methods requiring little skill and well known to all electricians".[52] Individual efforts were acknowledged, but not decisive. They found it "desirable that the determination of a quantity so important *should not be left in the hands of a single person*", even Weber, no matter how accurate his measurements were.[53] Discrepancies in determinations of the standard they attributed to the meager means – time, labor, and money – available to perform precision measurements alone. "In research of this kind", they explained, "the value of the cooperation secured by the committee of the Association is especially evident".[54]

The trials performed at King's College for five months during 1863 demonstrate the crucial role played by cooperation in reaching agreement over the precise value of the B.A. unit.[55] Significantly when these experiments were first mentioned in the reports of the committee, the point is made that over that period, "various sources of error [were] discovered and eliminated".[56] The deviation between the experimental results decreased over time, from 2.4% to 1.3% to 1.15%. Experiments were repeated again and again with the expectation that "a considerably closer approximation to the absolute unit" would be achieved.[57] In 1864 when the observed value of the standard reached 0.08% of its "true" value, the group decided to abandon the plan to create a perfect standard because it seemed not worth the effort to proceed any further.[58] Such collaboration led to synchrony, agreement, and even precision in measurement in the sense that their values clustered together within smaller and smaller ranges. The "true" value of the standard, as they had constructed it, was unity; the average of their composite mean values was 1.0001363. This first attempt was paradigmatic in the context of later determinations. The process demonstrated that for the British, the process of agreement, the negotiations concerning the precision measures representing the value of the standard, *took place inside the laboratory*. So convinced was the group of its personal agreement that any further *public* comment on the value of errors was not really regarded as so important. When, in 1864, the committee decided to stop measuring for the time being, they merely noted that the probable error of the result was "insignificant" and "will remain insignificant for any practical application".[59] In focussing on how their data was similar rather than what could still be done – that is, on how error could be further reduced – the committee was able to draw attention to the strengths, rather than the weaknesses, of its findings.

In the 1860s – and for that matter in the 1870s and 1880s, too – the Germans did not mount a collective or cooperative effort in determining a resistance standard. To be sure in Siemens's firm "laboratory", there was some minimal level of collaboration: Esselbach often did trials, Halske made instruments, and outsiders, such as Georg Quincke, sometimes provided corroborative support for Siemens's findings. But these efforts were part of a hierarchy of tasks, one might say, leading to a precision measurement. The Germans valued *individual labor* first and foremost. Even when Siemens discussed British work on the absolute resistance standard, he remarked that the "guarantee" of the "value" of the results was the reputation of the persons working on the project.[60] Overall Siemens argued for a different kind of practice in settling the value of the resistance unit. He wanted a concurrence of the results of many different observers using different methods; not until then, in fact, would he regard the B.A. unit as accurate to more than "some few per cent".[61] For Siemens, then, precision represented an agreement *between* individuals, not *among* them as a single collective result. Negotiations concerning the value of precision measures representing the standard took place *outside* the laboratory, in print. For this public negotiation over the value of the standard, the meticulous reporting of errors was crucial.

That is exactly what the British did not do. In the King's College experiments of 1863 and 1864, for instance, corrections for constant errors appear to have been adequately considered. Results were reported, however, in terms of final readings (two per day) and the mean of each set of daily readings. An assessment of individual values, their spread, and why they deviated from one another was omitted. The revelation that "the mean results of each day are more concordant than the individual experiments made on the same day", concealed more than it revealed.[62] Why were results from the same day discordant? Did constant errors still remain? In the production of precision measurements, the British were not working in terms of error and found little reason to focus on probable errors, much as they appear to have taken the time to calculate them. Hence even though a difference remained between the B.A. unit and the true value of the absolute unit of resistance, Matthiessen found the "close approximation" perfectly adequate.[63]

Siemens disagreed. "It is by no means satisfactorily proved", he argued, "that this professed agreement of the B.A. and the 10^7 meters/second units actually exists". Finding that, in the committee's 1864 report, differences between values amounted to as much as 9% and the mean values of pairs of observations differed by 1.4%, Siemens called the committee to task for its sloppy analysis of errors and cast doubt on the certainty of its findings.

By what train of reasoning the Subcommittee holds itself justified, in the face of such differences between even the means of their single observations, in concluding upon a probable error of only 0.1 per cent, I am totally unable to imagine. Whatever method is taken to calculate the mean values of given numbers, if some of the extreme measurements or some of the mean values be left out, very much greater differences are indicated.

In my opinion, certainty lies only within the limits of those numbers which are not liable to be regarded with mistrust as being faulty and therefore entitled to be cast out.

What Siemens viewed as the poor use of data analysis and probabilistic considerations by the British only intensified his suspicions of the cooperative effort responsible for creating the B.A. unit. It was impossible from the table of numbers provided, he pointed out,

to conclude upon a degree of exactness so great between the B.A. and 10^7 meters/second units, as is done by the Subcommittee; and when we recollect that the observations were made with the same apparatus, with the employment of the same correction-coefficients, the probability is considerably augmented that the difference between the two units is very much greater than it appears.

Ultimately Siemens suspected that the degree of exactitude achieved "was only attained in consequence of extraordinary manipulation and carefully selected methods of measurement". Siemens's suspiciousness about the B.A. result appears to have been a suspiciousness about the integrity of cooperative exact experimental work where the grounds for agreement on precision results were not adequately revealed. "It is very far from my intention", Siemens went on, "to assert that these measurements were not in reality made with all the exactness which is ascribed to them; they can, however, in that case only be the result of processes which have no general currency".[64] British exactitude was simply specious.

Siemens's criticisms jolted the British, but not in the way he had intended. Matthiessen missed the point about errors altogether.[65] So did Fleeming Jenkin, who also overlooked the function of Siemens's German silver copies and failed to appreciate the hierarchical orderings involved in German standards determinations. Objecting to Siemen's starting with a definition of a standard, and then allowing for practical improvements in its exactness, Jenkin clung to a standard's "invariability". Discrepant values were in his view an occasion for suspending judgment as to which value "is most near the truth".[66] Apparently convinced only by a narrow clustering of values, expressed as a percentage deviation between readings, Jenkin seems to have minimized the significance of error analysis – including the method of least squares – for understanding readings, much as it could have helped him. Error analysis could have helped him here, too. For Jenkin, the material conditions of an experiment, and of a standard, were more important than an analysis of the quality of the data either was responsible for. In his view the differences between the two sides constituted a gulf.

I know the immense difference between devising theories in the cabinet, or even trying isolated experiments, and actually carrying out those methods on a large scale by the aid of an organized staff. I concede both merits to Dr. Siemens; and if I have urged my arguments forcibly as to *the independence of the English school of electricians of that of Germany*, I beg Dr. Siemens to believe that I have done so . . . to vindicate myself from what I felt to be a very unmerited suspicion, that of having willfully omitted to mention his discoveries.[67]

Jenkin was right. What separated him and Siemens, the British and the Germans, was not a stubborn commitment to one type of metal or another, or a stubborn commitment to one level of accuracy over another. It was precision: how precision was achieved and defined, as well as the culture that encased it.

The differences between the two local cultures constituted a gulf. In one culture quantitative differences between the material and the real were eradicated as much as possible; in the other, that difference, as well as others, could be expressed quantitatively in terms of error analysis. One culture viewed data in terms of clustering; the other, in terms of ranges of error. For the Germans error analysis was essential for precise measuring and its practice. It was also needed, as we shall see, for the legal definition of a standard. As a result of these cultural differences, agreement on precison measures that would constitute a standard's definition was achieved in different ways. At a certain level the debate between the Germans and the British was not about a resistance unit, but about the labor practices and mental habits of precision. Until some common ground on precision and standards could be achieved, the B.A. unit and the Siemens unit might as well have been regarded as incommensurables.

THE CULTURAL AND SOCIAL FUNCTIONS OF ERROR

By the 1860s standards construction had become a complex social, political, and economic procedure. It was only a half century earlier that practices extending in some cases back to the middle ages had been superceded in a widespread weights and measures reform throughout the German states. At the beginning of the nineteenth century, in order to certify measures, it had not been uncommon to visit city halls, churches, and even libraries where embedded in the stone or plaster were two pieces of metal marking the end points of the local unit of length. Converting measures, necessary for trade, was difficult; for German weights and measures were marked by a particularism unmatched elsewhere. Accurate conversion tables were in demand following Prussia's 1794 *Allgemeines Landrecht*, which set new taxes requiring new cadastral maps (the new cadaster was a failure); the bureaucratic revolution of the late eighteenth century, which initiated a building boom; and the establish- ment of the French meter in the 1790s, which added one more measure to the dozens already converted in the markets of the German states. Before constructing his conversion tables for civil engineers in 1798 and again in 1810, the Berlin civil engineer and academician Johann Eytelwein journeyed to the traditional locations of metrological authority, checked the standards several times to vouch for their – and his – accuracy, and then carried out his conversions to five, six, and sometimes seven decimal places.[68] His results, carried out well beyond the limited accuracy of these older standards, were soon obsolete.

Weights and measures reforms were passed in Württemberg in 1806, Bavaria in 1809, and Austria and Prussia in 1816. Several western German states retained the meter imposed earlier by Napoleon. The *Zollverein* introduced a common weight, the half kilogram or *Pfund*, in 1839, which became legal in the post in 1851, in Prussian trade in 1856, and in coinage in 1857. But German standards remained diverse. The need for uniformity of weights and measures became especially apparent at the international exhibitions of the 1850s. At the same time in the Germanies, foresters, engineers, machinists, architects, and other representatives from industry, trades, and transportation spoke out in favor of establishing a single German system, preferably based on the meter. A pan-German commission composed of physicists, technicians, bureaucrats, civil engineers, and mechanics – without any Prussian representatives – finally met in Frankfurt in 1860 to decide upon a unitary system of weights and measures.[69] The commission's findings reveal the common basis to all German thinking – including, as we shall see, Prussian thinking – on systems of weights and measures.

The overriding concern of the commission was the foundation of the system it finally endorsed, the metric system. Neither the English nor the French, the commission found, had constructed satisfactory foundations. The commission found the major weakness of both systems to be the inaccuracies revealed by an assessment of the exact experimental protocols used to establish the standard. That the English yard as of 1855 no longer had a foundation in length of the simple seconds pendulum in London was less disturbing than the procedure used to measure the original yard. The English chose to measure between etched lines, rather than between end points. For this a microscope was required, but there was no sure guarantee that personal differences in reading from a microscope could be eradicated; inconsistencies in the "exact" measure of the yard would persist. There was, moreover, a crucial difference in convention. In assessing the temperature dependency of the length of the yard bar, the British did not specify the *error* the bar would suffer at a given temperature, but rather gave an expansion coefficient which could be used to calculate the temperature at which the bar would have the correct length. To the commission, what made any thought of "setting up the English foot as a foundation of the German system of measure completely alienating" was its finding that in England itself the system was in no way "unconditionally trustworthy". It would be a "great mistake", the commission concluded, to adopt the English system – or the methods used to realize it.[70]

The foundation of the French meter was not any more secure in the commission's view. The meter, originally set at 1/10,000,000 of an earth quadrant running through Paris, lost its "natural" foundation when later measurements of the earth proved the quadrant to be considerably longer. The commission cited Bessel's determination of the earth quadrant as 10,000,855 units of the original meter bar. Since the *Urmaass* in Paris did not conform to its original definition, the commission considered the meter an "*inaccurately determined measure*" in want of a more accurate foundation.[71] It did not thereby reject the meter; it merely rejected its foundation. (Doubts concerning the meter's foundation might have been a factor in German criticism of the B.A. unit, whose framers clung to the original earth quadrant measure even as the meter was being reassessed.)

There was both rhyme and reason to the commission's objections. Fleeming Jenkin was right when he said that when measuring, the French did not consider the earth's diameter or the British, pendulums; "for practical use, the material standard, not the definition, is the important point".[72] But in the Germanies, practical use was not the point in the legal definition of a standard, especially in the construction of *Urmaasse*. The commission's major proposal concerned the "manufacture of a German *Urmaass* for the meter". It held up as the methodological model for this important undertaking Friedrich Wilhelm Bessel's establishment and construction of the Prussian foot on the basis of simple seconds pendulum experiments in 1839.[73] The commission's decision is a testament to the widespread German consensus regarding Bessel's achievement and the significance it carried for subsequent standards determinations.

Bessel's work was the final stage in satisfying Prussia's 1816 weights and measures reform. Trials on the Berlin simple seconds pendulum were based on those done a decade earlier at Königsberg; in the interim, the Königsberg experiments had assumed the status of a paradigmatic exemplar of exact experimental methods.[74] With the assistance of the Berlin mechanic Thomas Baumann, Bessel constructed the *Urmaass* for the Prussian foot between 1835 and 1837, choosing to measure end to end rather than between engraved lines.[75] Few could duplicate what he and Baumann did, and for that reason the *Urmaass* could be regarded as inaccessible. Agreement on the value of the unit of length as expressed in the standard was not an issue: Bessel alone set that value, and his results were accepted without question or debate by the state. The state thus acknowledged Bessel's authority and sanctioned the use of complicated exact experimental methods in setting and validating a standard. Finally, Bessel made error analysis central not only in the scientific determination of the standard, but also in its legal definition. In his view an absolutely perfect standard could never be constructed; one had to know the range within which the value of the standard fell. The quality of a reproduction of the *Urmaass* depended on the value of its errors, which were also publically reported. For Bessel error analysis in the manufacture of standards guaranteed not only the quality of the standard and its copies, but also the honesty and integrity of the person producing them.

The Frankfurt commission recommended that Bessel's methods be those of a new committee, composed of at least one physicist and one mechanic, which would be responsible for constructing a German *Urmaass* for the meter. In electing to measure end to end rather than between engraved lines, the commission knew their recommendation would be subject to criticisms of how the metal of the bar and the end material (Bessel had used sapphire) changed differentially with temperature; but they thought that this could be analytically determined. It recommended, too, that copies of the *Urmaass* and *Normalmaasse* be made "which correspond to all the requirements of science". These copies themselves had a "scientific value", so they also recommended that several be placed in scientific institutes; among these would be a *Hauptnormal*, a main copy, that would remain stationary. Over and over again the committee emphasized that *all Normalmaasse* be made in the same *"Atelier"* so as to guarantee the consistency and the quality of all copies.[76] The committee's recommendations thus reinforced the highly centralized, hierarchical, and controlled production of the *Urmaass* and its copies that had characterized Bessel's work on the Prussian foot.

But as it had been for Bessel, for the Frankfurt commission the real public guarantee of the quality of the standard or its copies were their errors, which the commission expressed in terms of "the tolerance [*Toleranz*] of weights and measures". A weight or measure, the commission argued, could not correspond "perfectly" to its legal value [*Sollgrösse*] for "technical reasons" and therefore one had to be satisfied with "a certain degree of approximation" to this legal value. But what should that approximation be? "Through an accuracy driven too high", it acknowledged, "the cost of testing will be unnecessarily raised and the desired goal [i.e., creating a copy] will therefore frequently not be reached". But some degree of accuracy had to be specified, much as "directions" about how to compute that accuracy might appear "superficial" to those unaccustomed to such calculations.[77]

To make its point, the commission explained the linkage between error in the laboratory and crime in the marketplace. Certain conditions, such as temperature, the commission explained, affected the accuracy of a copy. Disturbances due to the "effect of natural forces" could be quantified by error analysis, just as Bessel had done for the Prussian foot *Urmaass*. Repeated manufacture of copies provided information on the degree of accuracy that could reasonably be expected from any copy; this degree would be made the legal tolerance. But whether or not measurements would in fact fall within the range of tolerance ultimately depended on human performance. "Above all else", the commission pointed out, "the greater or lesser carefulness with which the measuring or weighing is carried out" influenced accuracy. In the commission's view, "uniform practices" in weights and measures existed only when "determinations fell within the range of admissible tolerance". So ideally, "control of the remaining influences dependent upon the will of the people" was needed to obtain results that "deviated from the truth as little as possible;" that is, protocols for using weights and measures had to be specified so that all readings fell within the computed range of tolerance.[78] Deviations in certain directions outside that range could be considered grounds for suspecting illegal activity. At all times the concept of legality and illegality in weights and measures was predicated on the prior accurate determination of the range of tolerance of the copy, and this required an extensive analysis of errors.

In daily use, however, the concept of tolerance was not monolithically defined and hence the legal/illegal distinction varied from circumstance to circumstance. In commerce and trade, for instance, any weight or measure that fell short of the legal value had to be discarded because it was the buyer who suffered through their use. In biblical fashion, however, weights and measures that exceeded their legal value were admissible (buyers received more for their money), but the excess could amount to no more than twice the allowable deviation. And in general the repeated use of weights and measures in trade and commerce was acknowledged to produce deviations that were inadmissible in more controlled settings. By contrast, the various kinds of *Normalmaasse* – the *Hauptnormal*, *Kontrolnormal*, *Gebrauchsnormal*, and the *Normalmaasse* kept at the regional and local offices of the bureau of weights and measures – were to be more accurate, have fewer errors, and smaller ranges of tolerance. Consistent with the demands of both Weber and Siemens regarding the need to align the accuracy of standards with precision of sensitive instruments already in use, the Frankfurt commission even considered precision measurements in framing the concept of tolerance.[79] Weights and measures used for scientific purposes and weights used for precious metals and coins were constrained to smaller ranges of tolerance. But there was a price to pay: the smaller the range of tolerance, the greater the control – technical and behavioral – that had to be exercised to ensure that economic exchange conformed to the legal dimensions of weights and measures.

German *Urmaasse* were constructed by the decade's end according to the guidelines set out by the Frankfurt commission. When the North German Confederation adopted the metric system on 17 August 1868, it did so on the basis of a German *Ur*meter calibrated to be 1.00000301 of the original French meter and a German *Ur*kilogram equal to 0.999999482 of the original kilogram. In ways that had not been done before, the Confederation edict enhanced the function and the authority of local weights and measures offices by requiring them to test all weights, measures, and related apparatus to validate their accuracy and then to stamp those that yielded values within the legal ranges of tolerance. Daily life would never be the same. Even wine bottles, set at one liter capacity, were subject to testing and stamping. But beer, "the most important drink of the people", was left unregulated except for its alcohol content.[80] As later commentators recognized, stamping apparatus was an admission that mathematical certainty in weights and measures operations was impossible; that determining ranges of error and setting levels of tolerance was a way to control but not eradicate uncertainty; and that insofar as "correct handling" was defined for using weights, measures, and related apparatus, even human behavior had to be controlled to obtain accurate results.[81] Stamping constrained error to remain within certain limits; but it was also an acknowledgement that deviation could be held in check but never completely eliminated.

Error was thus a key concept in German weights and measures determinations and operations. It had several functions: scientific, in exact experiment where it indicated the credibility of a result; metrological, in defining the precision of a standard; technical, in the concept of tolerance; ethical, as a sign of the integrity of the investigator or of the seller; commercial, in the determination of a standard's value; and above all, legal, in defining where criminal action began. Only with error and the concept of tolerance were there adequate grounds for contesting a measurement; only with error and tolerance, therefore, could illegal activity in the marketplace be identified. Error thus shaped the social meaning of precision measurements in German weights and measures. Quite simply, without error there could be no standard.

REACHING CONSENSUS ON CONVERSION VALUES

Weber, in contact with William Thomson on the matter of constructing an absolute unit of resistance, drew upon the Frankfurt commission's findings when he returned to the issue of a resistance unit in the early 1860s. His recommendations incorporated distinctly German traditions in weights and measures. Admitting that his own earlier measurements on the absolute value of the resistance unit were never intended as definitive, he argued in 1861 that if his absolute resistance unit were chosen as the legal unit, then it would require the same kind of foundation that Bessel had given for the *Urmaass* of the Prussian foot through the seconds pendulum experiments. "It would be necessary", Weber explained, "to create an absolute resistance measure according to the most rigorous rules, with the most perfect instruments, and with all the devices of the finest observations. This is a problem which can only be solved by very skilled hands, with large blocks of uninterrupted spare time, and in better institutions than are presently available for physical research".[82] He considered, however, "the most pressing need" at the time of his writing to be the construction of a reliable and accurate standard for telegraphy, so he turned his attention first to the construction of copies, or *Etalons*, for public use. Their identical manufacture had to be unconditionally guaranteed.

Just as one can regard many exemplars of a text as identical only when they are made with the same press, one can achieve *Etalon-exemplars* regarded as identical not as a result of artistic or handmade copies, but only through *stereotypical impressions* made with the finest and most reliable machines under conditions that remain completely unchanged. Not the stamp of the office of weights and measures, but the stamp of the machine used to make all examplars guarantees in copies an equal surety suited for practice.[83]

Such machines, he acknowledged, did not yet exist; but a first step toward the production of exactly identical *Etalons* could be taken by improving certain measuring instruments, particularly the galvanoscope, used in the manufacture of *Etalon* resistances. Hence his creation in 1862 of *Galvanometrie*, the science of resistance measuring, in which error analysis and tolerance played a crucial role; for in Weber's view, "copying rests on a judgment about the equality or inequality of two values".[84]

For those directly involved with practical problems in telegraphy in the 1860s, especially at various branches of the Siemens firm, the science of resistance measuring involved persistently redesigning fairly traditional apparatus for the purpose of attaining higher accuracy and more convenient handling.[85] At the same time that specifications for the construction of *Urmaasse* and for the manufacture of copies became more intricate and exacting, resistance measuring in general was simplified considerably through the efforts of those with knowledge of what was required in the field, both in terms of human performance and in terms of problems to be solved, especially fault location.[86] The gap between the labor that went into the construction of a standard and its practical use widened considerably when instruments for "inexperienced observers", capable of high degrees of exactness, were created.[87] The repertoire of instruments designed for the telegraph engineer eventually complemented those of the physicist and possessed corresponding degrees of exactitude.[88] In announcing one of the most important instrumental breakthroughs of the decade, the universal galvanometer, Werner Siemens himself explained in 1868 that measuring resistance, current, and electromotive force was "no longer carried out exclusively by *Physiker von Fach*, but also by telegraph engineers, and far more frequently".[89] His universal galvanometer was designed to streamline and simplify the telegraph engineer's tools by combining in one instrument the devices needed to measure all three quantities. Such instrumental improvements increased the accuracy of practical resistance determinations in the 1860s and 1870s. But other factors drove German investigators into a deeper examination of the precision of the Siemens unit: more rigorous weights and measures specifications, including the assessment of error; the continuing debates with the British, especially concerning consistency in the reproduction of the Siemens unit; and finally the official adoption, at the 1868 International Telegraph Conference in Vienna, of the Siemens unit for all international business. As Franz Dehms explained in 1868, "the increased requirements for the accuracy of the *Normalmaass*, as well as objections which English scientists, especially Herr Matthiessen, have raised against the mercury unit, have led to a repetition of work on the unit in Siemen's laboratory". At Siemens's request, Dehms remeasured the Siemens resistance unit, attempting to bring its accuracy in line with the redetermined *Normalmaasse* for length. When Dehms remeasured the length of the mercury tube, for instance, he sought the assistance of Adolph Brix, Phillip Wilhelm Brix's uncle and head of Prussia's bureau of weights and measures, who made available a recalibrated *Normalmaass* for Dehms's trials. Almost the entirety of Dehms's investigation was taken up with an analysis of the errors in his determination as he tried to counter, with strict quantitative reasoning, each objection raised against the accuracy of the Siemens unit. In the end he was able to claim a difference between observed and calculated values of 0.05%. "Such an agreement, which one may deem near perfect", Dehms concluded, "eliminates the final objections against the reproducibility of the mercury unit. It also speaks enthusiastically for the superiority of the present form of the measuring instruments and of the way in which they were handled".[90]

As an afterthought, Dehms decided to express the B.A. unit, the ohmad, in terms of the Siemens unit. Conversions between these two units, and others, had been at issue since the British Association published a conversion chart of "Approximate Relative Values of Various Units of Electrical Resistance" in 1864.[91] The 1864 B.A. conversion value was 1 B.A. unit for every 1.0456 Siemens units; Dehms found the value to be 1.0493.[92] In its emphasis upon errors, its reliance upon weights and measures specifications, and its execution by a solitary investigator, Dehms's experiment exemplified what soon became the German route to consensus on the conversion values of the two resistances. The computation of conversion values based on more accurate determinations of the Siemens unit also became the principal way of raising objections concerning the value of the ohmad and of the manner in which it had been determined. Over the course of the next decade, the most famous objections were those raised by Friedrich Kohlrausch, Wilhelm Weber's former assistant in the physical practicum at Göttingen.

The Göttingen program in physics was originally based on Weber's and Gauss's earth magnetic measurements, but over time it embraced the measurement of all electrical quantities as well as rigorous training in the quantitative analysis of instruments and the constant errors of an experiment. Early on the determinations of the horizontal intensity of the earth's magnetism assumed the status of a paradigmatic measurement, useful for teaching students the central components of physical measurement in much the same way that Bessel's seconds pendulum investigation did at Königsberg.[93] It was a difficult measurement, fraught with possible errors. When Weber proposed measuring resistance in terms of absolute units, his methods included a consideration of the horizontal intensity of the earth's magnetism. The British, aware of the difficulties involved, tried to develop alternative methods that would bypass it. Kohlrausch did not. Recent developments in the British Association, he remarked in 1870, made it necessary "to determine with the highest possible accuracy the relation of the Siemens unit to the Weber absolute unit [of resistance]". He especially took the British to task for their laxness in reporting data, which foreclosed any independent judgment of the data's irregularities and the causes of that irregularity. Although he praised the theory behind the corrections they offered and the elegance of their report, he suspected that additional errors still to be discovered lurked behind the scenes, especially the error caused by uncertainties in their measure of the earth's magnetism. That other errors might be present made it unlikely, he believed, that the British measurements had a probable error of only 0.1%.[94]

Four years later, after leaving Göttingen, he published his completed findings containing the unwelcome news that the B.A. unit was inaccurate by 2%, an error in all likelihood due, he argued, to insufficient consideration of the effect of the earth's magnetism on a multiplier needle set in rapid and uniform rotation. Kohlrausch's own determinations were done with his student assistant, W. A. Nippoldt, at the Göttingen magnetic observatory in 1869, using mercury units supplied by Siemens. Kohlrausch spared no praise for his own qualifications – and for the suitability of the Göttingen observatory – for carrying out the experiment. Individual skill, in his view, certified the certainty of his results. "I have shunned no labor", he wrote, "in order to reach the degree of accuracy that is attainable with such means; and in assuming that this goal has actually been reached, I believe it is guaranteed by years of practice in observations of this nature, no less than by the final agreement of the result". But his arrogance was mixed with humility, with an openness to criticism, and with an acknowledgement that assessments regarding precision measurements were essentially public negotiations.[95]

In Kohlrausch's view, others were supposed to judge both his results and his honesty not only on the basis of his experience – although that counted heavily – but also, and perhaps to a greater degree, on the basis of how well he reported on and analyzed his data, something he felt the British had neglected. He communicated his method and his results so "that it may be possible to form an opinion as to the degree of accuracy that has been reached" because "in the case of any fundamental and difficult measurement which stakes a claim to exactitude [Exactheit], the fullest detail is desirable in the publication of the numbers obtained". The British, he felt, had not published the "full details of a series of observations" – a weakness in an otherwise classical report – and so judgments concerning the accuracy of their results could not be made. He was suspicious, too, of what appeared to be an inability on the part of the British to distingish constant and accidental [random] errors and to judge where constant errors ended and accidental ones began. He warned the British that "*in measuring physics it is always dangerous to assume that large experimental errors can only be of accidental origin and are eliminated by a sufficiently great number of observations*". His own examination of the limited data they provided demonstrated that there were certain patterns in the deviations which might be explained by deeper investigation of the constant errors of the experiment, and he named several possible ones. He felt every "unprejudiced reader" should have their "attention arrested" by irregularities in the data that they should then seek to explain. "In short", he concluded, "it does not appear to me allowable to apply the rules for the calculation of probabilities to such observations unless we have explicit proof of the absence of constant sources of error".[96]

Although he computed the probable error of his own results as 0.05%, he did not consider his findings conclusive, but simply as grounds for further negotiations.

I by no means maintain that with such a limit of error *all* that is desirable is accomplished; but without constructing entirely new instruments, and without erecting special buildings, it would be for the present difficult to exceed this limit.[97]

It had not been the purpose of his investigation to decide between the two different resistance units; but he believed in the end that the Siemens unit had the advantage with its resistance scales, accuracy of definition, and ease of practical application. He computed a final value for the Siemens unit as 0.9717 earth-quadrant/second, or 9,717,000,000 mm/sec (or, 1 ohmad = 1.0291 Siemens units). In comparing his result to "the most trustworthy value hitherto published", that of Dehms, he concluded that the British had erred by 2% in the value they assigned to the ohmad.[98] That Henry Rowland found an error in Kohlrausch's equations[99] and that Kohlrausch's value for the ohmad turned out to be among the lowest computed internationally does not detract from the significance his investigation holds in the history of the German determination of the unit of resistance. It reinforced the traditional German pattern in the treatment of standards: by way of individual investigations steeped in error analysis, with special importance assigned to probable errors.

One sign that conversion results were meaningful to more than those who calculated them can be found in textbook and handbook culture of the 1870s. In his handbook on electricity for telegraphers of 1876 the Siemens physicist Oskar Frölich remarked that with the construction of standards and resistance scales, resistance could be determined in much the same way, and with a corresponding accuracy, that physicists determined weight. Given his experience at Siemens and the sanction given to the Siemens unit at the 1868 International Telegraph Congress in Vienna, Frölich saw no need to explain the ohmad – now also called the ohm – in detail. The loss of a "natural" foundation to the meter and its recalibration had by then raised questions about the foundation of the B.A. unit; the meter's changing value with respect to an earth-quadrant measure meant corresponding changes in the value of the ohm. Moreover, in his view, the ohm was more difficult to determine than the Siemens unit. The value of the Siemens unit possessed greater certainty given the relatively higher purity of mercury compared to other metals and given that the unit's value was based on measures of length and specific gravity, which physicists, benefitting from the results of weights and measures reform, now could determine more accurately. Still, he could not dismiss the ohm; for he detected changes in electrical technology requiring different metrological considerations. The Siemens unit sufficed for telegraphy, he argued, where the most important problem was still fault location – he called it "the best test of the skill, the experience, and the insight of the electrician". But when problems concerning heat and work arose, "the engineer requires certain fundamental determinations which treat the relation of various forces to one another", and for this the system of absolute units was most suited. Frölich had thus reached the position that had been William Thomson's point of departure years earlier. So in closing, Frölich introduced the absolute system to practicing electricians and expressed the Siemens unit in terms of ohms, for which he used Kohlrausch's 1870 value.[100]

Spurred by changes in instrumentation and electrical technology, the physical deterioration of earlier resistance standards, weights and measures reforms, and in 1881, the inauguration of International Electrical Congresses, German physicists redetermined the Siemens unit and its conversion value for the next two decades.[101] Especially following the 1881 Congress, where a call was issued for the precise length of a mercury unit equivalent to one ohm, international interest in resistance determination soared. One of the first major German reviews of resistance determinations, by Gustav Wiedemann, made it clear that negotiations on the value of the Siemens unit were still to take place publically through a discussion of "the methods used and the sources of error". Wiedemann even took Henry Rowland, Kohlrausch's critic, to task for his inadequate data analysis. Rowland apparently carried out his results to so many decimal places that Wiedemann believed that they could "only be regarded as interpolations". Hence, Rowland's accuracy was "one-sided" and it could not "guarantee the certainty of the final result".[102] Employees at Siemens & Halske cast the "negotiation through error" dictum in their own terms: manufacturers of standards, including resistance standards, had an obligation to make known their "manner of justification" so that it could be publically discussed.[103] A new element entered arguments concerning the precision of resistance determinations. In addition to the customary evaluation of precision through application of the method of least squares, these investigators used official *Normalmaasse* for determining length and weight in their trials; a third generation copy or *Etalon* was just not good enough any longer to certify the accuracy of a result. In these length and weight measurements official ranges of tolerance was crucial for judging how many decimal places were reliable.[104] The earlier reform of weights and measures thus contributed measurably to the more accurate determination of the Siemens unit and its conversion value.

In these later German resistance determinations, there is some evidence that the organization of labor in precision measuring became more complex, moving cautiously beyond the individual laborer. Resistance determinations at Siemens & Halske, for instance, often depended upon at least a two man team, a physicist–electrician and a mechanic, and sometimes more. The report on the firm's 1882 determination of the mercury unit was written in the third personal plural; the author was listed simply as "Siemens & Halske", indicating that several individuals contributed to this difficult project.[105] Judging from Wilhelm Weber's final measure of the absolute unit of resistance, undertaken with Friedrich Zöllner at Leipzig in 1880, the increase of scale and complexity in precision measuring extended beyond the labor required to do it. Weber and Zöllner used a team of four additional individuals for trials and data processing. Work on the apparatus involved others at Repsold & Söhne in Hamburg and Siemens & Halske in Berlin. Their apparatus and copper wire alone cost over 7,000 marks. The team recognized that their investigation was not only a matter of constructing a "normal conductor" with a high degree of accuracy, but also of making the observations simple so that they did not take up too much time.[106] Precision measuring was beginning to look like a problem in the cost-effective management of resources.

Such practices, however, did not yet become widespread, nor did they significantly alter deep seated assumptions about how agreements were to be achieved regarding the value of precision measurements, that is, as public negotiations based on error. Where more than one person was involved, the organization of labor was hierarchical, so although the effort was collective, it was not necessarily cooperative or collaborative. Decisions concerning the final result still generally fell into the hands of one person. German investigators continued to place a high value on *individual* effort. So when Wiedemann, in his review article, called resistance measuring "a rich field for accurate investigations" for which there were, in 1882, only "a series of pre-investigations", he assumed that future investigations would still be conducted by individuals according to their own plans. "The most reliable final result", – that is, agreement on the value of resistance – would then be found through a comparison of the data of many investigators.[107] The working assumption behind this endorsement of individual effort publically discussed was that a consensus result could not be based on "one method being carried out in one place". A result could be protected from error only "if it is derived out of a series of trials completely independent of one another with different kinds of apparatus".[108] In a similar fashion, Siemens & Halske argued for the publication and discussion of methods of justication as the means for achieving agreement. Every instrument the firm calibrated, they pointed out, was accompanied a statement of justification to serve this purpose.[109]

The British, too, urged experimental variation in deciding upon the value of the ohm. "It is only by the concurrence of evidence of various kinds and from various sources", Lord Rayleigh explained to the B.A.A.S. in 1882, "that practical certainty may at least be attained and complete confidence justified".[110] But the British position, as Simon Schaffer has argued, was that precision measurement "hinged on communal trust"; that precision values had a "collective quality"; and finally that "precision measurement must be collaborative".[111] As Maxwell and Jenkin pointed out in 1865, "exact knowledge is founded on the comparison of one quantity with another", and for that purpose absolute units were well suited. But they assumed a particular *process* of comparison. "In many experimental researches conducted by single individuals", they emphasized, "the absolute values of those quantities are of no importance; but *whenever many persons are to act together*, it is necessary that they should have a common understanding of the measures to be employed". In Britain, then, the comparison of precision values was a collaborative social enterprise; in Germany, a competitive one based on individual effort.[113]

The German approach had its consequences. David Cahan has observed that members of the German delegation to the first International Electrical Congress, in 1881, "were concerned about [the Congress's] ability to reach trustworthy decisions".[114] Larry Lagerstrom found that, in a similar vein, delegates to the 1884 Exhibition could identify "no criterion with which to sift the accurate results from the inaccurate", due partly to the "variety of methods" that had been used and "the inability to examine the details of each calibration".[115] The complications of striking an international agreement aside, these results should not surprise. Several key representatives at these congresses – Siemens, Wiedemann, Helmholtz, and Kohlrausch in the German delegation, H.F. Weber for Switzerland, and Heinrich Wild for Russia – were accustomed to a negotiating style based on a discussion of errors both in precision measurements and in electrical standards determinations. As Ernst Dorn explained about the measuring apparatus and data on exhibit at the 1882 International Electrical Exhibition in Munich, the exhibit was designed to allow viewers the opportunity to make their own "independent criticism of all the data".[116]

That is not to say the German delegates to these congresses did not want to work toward a common goal. The Germans were in fact the ones, who, at the 1884 Paris Congress, made the recommendation that they all agree on where to disagree by determining the value of the ohm to at least 1/1000 of its true value. In the meantime, the proposed legal value of the ohm was set at 1.06 Siemens units, which the Germans felt

comes so close to the theoretical ohm that for all technical purposes . . . the difference can be neglected. On the other hand, for scientific investigations which strive for an accuracy of greater than $\frac{1}{2}$ percent, at most a small correction will be necessary, whose value will be determined through more accurate investigations in the future.[117]

Although practical and scientific problems remained – decisions still had to be made on how *Etalons* were to be manufactured – German representatives recommended that their government accept this legal ohm as the standard unit for electrical resistance.[118] But despite their arguments concerning the need for a standard resistance for legal contractual purposes and generally for those areas of "electrical technology set in motion by significant capital", the

German government did not immediately act on the matter.

At the national level, the British moved more quickly than the Germans: their 1889 weights and measures act included a section on electrical standards; they reviewed the value of the ohm in Siemens units in 1889; and in 1891 they set the ohm at 1.063, but recommended a metal rather than mercury standard.[119] In Germany, by contrast, where Siemens believed there was no place suitable for carrying out the difficult measurements for establishing resistance units, arguments for a proposed institute for precision measurement continued to value individual over cooperative and collaborative effort. A commission on the matter at the *Akademie der Wissenschaften* argued that investigators

even if they have to work in their own locality under somewhat unsuitable conditions, would achieve more reliable results than could be had at an international institute built by European governments because they would try to carry out their own plans and discoveries more than they would be able to as officials at such an institute where the method of execution would necessarily be prescribed. Prescribing a method is dangerous. The reliability of results in the present state of [electrical standards determinations] will be best controlled by the agreement of all possible different methods.

It is not that state support was not desirable; for the commission acknowledged that precision work of this type required a greater expenditure "than most physicists can draw upon from private means or from the funds given to them for instructional purposes". Without abandoning the commitment to individual investigations, the *Akademie* commission argued that state funding was absolutely essential for a "physical observatory", as they called it in 1883, *"even if it will come too late for the determination of the galvanic resistance unit."*[120] Invoking practices in weights and measures, the commission acknowledged in 1885 that the legal establishment of the resistance unit "would not be possible if a place were not established for the official testing and certification of resistance *Etalons* and instruments for measuring current". The result in 1887, of course, was the Physikalisch-Technische Reichsanstalt, whose story has been told in admirable detail by David Cahan.[122] The PTR's role in the legal and scientific establishment of the ohm eventually proved decisive.[123] But traditional ways of establishing a standard and reaching consensus on precision measures, however, were not entirely abandoned.

THE GERMAN SETTLEMENT

The establishment and testing of electrical units and standards, including the construction of *Normalmaasse* for resistance, were part of the reponsibilities of the scientific and technical sections of the PTR.[124] Members in both sections worked on the value of the ohm with facilities hitherto unavailable in Germany but in an environment that David Cahan has described as a "rule-governed, bureaucratic system of scientific organization [with] a hierarchical work structure that consisted largely of small teams, not individuals".[125] Yet when PTR director Hermann von Helmholtz wanted a rigorous evaluation of the value of the ohm, he significantly appointed not one of the PTR's teams, but an outside commission whose members worked independently to determine "the most probable ratio of the theoretical ohm to the Siemens unit with the greatest possible clarity".[126] Helmholtz chose as the commission's members the physicists Gustav Wiedemann, Friedrich Kohlrausch, Franz Himstedt, and Ernst Dorn. Although all four came to the same conclusion – the most probable value for the ratio was 1.0628 – Helmholtz singled out Dorn's report for publication with the PTR's recommendation regarding the ohm. The choice is revealing. Dorn was known for working in the exact experimental style he had learned between 1866 and 1869 in Franz Neumann's Königsberg seminar for physics where error analysis, especially the determination of probable errors, played a central role and at times even became the raison d'être for a physical investigation. While a secondary school physics teacher in Königsberg in 1872 he published reports on measurements taken at a new geothermal station; in it he focused on theory, reliability, and calibration of the station's thermometers with intent of improving upon Neumann's method for thermometer calibration. He began electrical measurements while *ausserordentlicher* professor of mathematical physics at Breslau between 1873 and 1881 and continued them as *ordentlicher* professor at the Technische Hochschule in Darmstadt in 1882 and then at Halle in 1886.[127] These latter investigations caught Helmholtz's attention.

The request made at the 1881 International Electrical Congress in Paris for the conversion value of the Siemens and ohm units prompted Dorn to publish his first set of investigations on the reduction of the Siemens unit to absolute measure. Dorn explicitly presented his findings "in a form which permits a judgment of the method and its execution". So he spared his readers no detail. Like Kohlrausch, he attributed the main source of error to measurements of the horizontal component of the earth's magnetism. And like Kohlrausch, he spared no praise for his skill in carrying out the investigation. Dorn carried out the measurements "with all desirable sharpness"; he offered controls for his measurements; and he especially "spared no effort in the arrangement and the reduction of the observations". For measurements of the mercury's weight and the tube's length – two crucial parameters of the Siemens unit – Dorn used *Hauptnormale* from the Breslau office of the bureau of weights and measures. Convincing readers of the certainty of his findings extended beyond their independent judgments of his measurements and conclusions; for Dorn wrote his report in such a way to make readers feel as if they were present during the investigation. In minute detail he described the location where the measurements were taken: in a large one-story high room in the "Hostel Building", now a part of Breslau University but formerly belonging to a monastery, possessing very thick walls and vaulted passageways. Measurements, he assured, were only taken on university vacations, but at the end of the day, between 9 p.m. and 1 a.m., to avoid the vibrations caused by street traffic. He admitted that he had to take other measurements during the day, but only on Sundays while church services were in session. Complementing the visual detail of his experimental set-up was the analytic detail of his report on his data. He gave a complete listing not only of all measurements, but also of the data and time they were taken and their probable errors. Dorn reported a final value of 1 Siemens unit equal to 0.9482×10^{10} mm/sec [Siemens unit/ohm ratio: 1.0546].[128]

Dorn received assistance from Frölich and the Breslau University physicist Oskar Emil Meyer – both also graduates of the Königsberg seminar – and from Siemens & Halske, which supplied some of his instruments, for this first major study on the value of Siemens unit. But as others before him had found, increasing precision sometimes meant investigating on a larger scale. In his second major study on the Siemens unit, Dorn's precision measuring operation became even more complex: Kohlrausch, Frölich, and Meyer assisted from time to time, and for taking and reducing data, Dorn employed the help of two recent graduates, five students, and a mechanic. This time rather than using Normalmaasse located in Breslau, Dorn borrowed a set of weights and a meter bar from Johann Pernet, then working at the Bureau international des poids et mesures in Sèvres, France; Pernet, who also attended the Königsberg physics seminar, had calibrated the meter bar himself. Dorn also had to rely on others to produce the various size resistances he tested. Dorn was, however, responsible for designing, supervising, and carrying out the major determinations in the experiment. This time, to reduce errors still further, he abandoned methods that required measuring the earth's magnetism. And once again he described the location of his experiment: a room in the Darmstadt Technische Hochschule from which all the iron was removed. But now to make the setting of the experiment more immediate to his readers, Dorn included a drawing of the room and its apparatus. Again he took his data primarily in the evening; but now, cognizant of the need to control temperature, he allowed only two persons and two candles in the room at one time. He provided the same meticulous detail on the constant errors and corrections they entailed, his data, the dates and times they were taken, and their reduction. The Siemens unit/ ohm ratio this time was 1.0624.[129] Through detail, Dorn intended to persuade.

Dorn's determination of the ohm's most probable value was completed on 16 October 1892, two months after Helmholtz negotiated an agreement at the 1892 International Electrical Congress in Edinburgh that the ohm would be represented by a mercury standard in absolute units. Despite the international agreement, details on the value of the ohm were worked out nationally. Conducting no new investigation of his own, Dorn reviewed all prior determinations of the absolute value of the Siemens unit for the purpose of conducting a more extensive analysis of errors than the authors had originally given, and on that basis, recomputing their results, including their probable errors. He redid none of the trials; he inspected no apparatus. His critical assessment was based entirely on quantitative operations that he could perform on paper. He computed the mean in two ways. When he treated all redeterminations equally, the mean was 1.06274 ± 0.00023. But he clearly did not trust all results equally – he had even thrown Heinrich Wild's results out because Wild had reported insufficiently on the details of his experiment – so he also recomputed the mean by assigning half weight to determinations that appeared less secure. The result this time was 1.06289 ± 0.00024. Dorn finally concluded:

It is difficult to decide which value should be adjudged as having the greater probability. One may, for the time being, consider 1.0628 as coming very close to the true value. If it is a matter of choosing between 1.062 and 1.063, however, preference can be given to 1.063.

The overall agreement that Dorn saw in his data became for him a "guarantee" that the "authoritative laws of nature" in this area were "sufficiently known", a conclusion that he considered "likely" even in the context of "the wider standpoint opened to us by Hertz".[130]

Dorn's investigation was both a confirmation of, and a departure from, traditional German methods for establishing the value of a standard. Like standard determinations of old, Dorn's determination of the most probable value of the ohm was an individual investigation; for in essence he had simply normalized prior determinations according to his own criteria for analyzing errors and data. A recomputation of constant errors put his mark on each experiment. The results were no longer those of the authors; they were Dorn's. Essentially Dorn created a single observer out of many sets of measurements. Read from a traditional point of view, Dorn's study reinforced the central role of an exact experimental physics, with its rigorous error analysis, in standards determinations. And from a traditional perspective, the "true" value was that based on the differentially weighted data, 1.06289 ± 0.00024. Precision, and in this case, also the standard, were both defined through error. Negotiations on this value took place only in the sense that errors were compared. One sought "agreement" by considering overlapping spreads, each determined by the probable error of individually determined averages. From a traditional point of view, the result based on differentially weighted data would have been the scientific and the legal value of the ohm.

But that is not what happened. One could in fact consider Dorn's study a collaborative determination – several individuals did contribute, and near equally – without any real group discussion. The experiments and the data remained in some sense "different" among themselves. One could then view the three sets of values Dorn provided – two weighted values, a near-average of 1.0628, and a "rounded up" value of 1.063 – as levels of compromise in reaching agreement. The PTR's curatorium recommended in 1893 that the last value be integrated into law, and it was in 1898. The decision was based on (1) the discovery of another error, not considered by Dorn, stemming from the layer of air remaining between the mercury and the tube;[131] and (2) practical considerations (using it or 1.0628 would make no difference in applications). This decision in effect constituted a relaxation of former prescripts for precision; for despite corrections stemming from the additional error, the last decimal was still dropped. "Old-style" precision did not, in this case, a standard make.

By discarding the more "traditionally" accurate determination, one could also say that the meaning of precision changed. In the traditional view, the precision of a result was defined by its range of error. But in this instance, "rounding up" suggested that it was not error that was focused on, but rather the point around which most actual readings would cluster, much in the same way the British had done. Dorn had reviewed prior investigations using traditional standards, but once he created three sets of results and recommended either of the two last, it was really clustering that he sought to achieve. The British had likewise thought of precision in terms of this clustering of values; for they had made consistent and greater reference to simple percentage deviations from the mean result rather than to probable errors. Negotiations leading to agreement focused on this clustering rather than on error and data analysis. So when the German Reich legalized the ohm in 1898, precision – in the sense of a rigorous adherence to the results of exact experiment, including the calculation of errors – was relaxed somewhat in favor of expanding the community of individuals who could agree without objection on a numerical result. In choosing a mercury unit of 1.063 meters length, framers of Germany's electrical units law, which included Dorn and members of the Physikalisch-Technische Reichsanstalt's curatorium, put aside the promise of high degrees of precision which had driven investigations hitherto.

An absolute measurement of a physical quantity can naturally never be achieved with an accuracy greater than that which is attainable through the manufacture of the unit of measure for the quantity in question. *The greatest certainty and agreement* in the measurements of various observers can only be reached therefore, if in the manufacture of units of measure, *those methods are used which make possible the most accurate agreement.*[132]

Reconsidering Dorn's value in 1898, framers of the law argued that the second decimal was "in no way guaranteed", so "for a lawful determination" the value could be "restricted to the first decimal" and set at 106.3 cm.[133] The likelihood of increasing agreement – that is, of creating a dense and narrow cluster of measurements made by several observers – prevailed over a strictly rigorous exactitude that had viewed a standard's precision in terms of the small ranges of error achieved by one highly skilled observer or at most a few. The settlement that defined the ohm thus emerged with a different concept of precision, one that did not, in negotiations, focus so strongly on error.

There are ambiguities in the way Dorn's efforts could be viewed, but what is clear, is that a threshhold had been crossed: precision practices after his determination of the resistance standard deemphasized error in favor of agreement and consensus. Never before had the value of a standard been established by "rounding up" and by so strongly diminishing the role of error. The decision to do so had its consequences. The law for electrical units generally conformed to German practices in weights and measures: how the various *Normale* would be constructed, the role of the PTR in testing and certification, and so on, were adequately discussed. But closure on the issues of crime and punishment, which depended upon a rigorous determination of error, was lacking. The range of tolerance was left unspecified. Instead, the PTR was supposed to determine "tolerable deviations" in measuring apparatus, but it was unclear how policing deviations would be done in public without a prior rigorous determination of error. Strangely enough, punishment was specified – for the use of incorrect apparatus one could receive a fine of up to

100 marks or up to four weeks in jail and the measuring apparatus *might* be confiscated – even before the crimes themselves were defined. Merely to say, though, as the law did, that "the use of incorrect apparatus is forbidden" was grossly insufficient when, again without legal ranges of tolerance, stamping of apparatus could not begin. Assuming stamping could be done, the framers of the law were not even clear on how to proceed in cases of crime in the marketplace where electricity was traded. Sellers might have knowledge of the measuring apparatus, but these framers were not clear on the matter; buyers definitely were not liable in cases of fraud because the framers assumed they would not have knowledge of the apparatus and hence would not know how to cheat. Moreover these framers worried that seizing incorrect apparatus – done often enough in the commercial marketplace – might not be "harsher for those involved than the punishment itself".[134] In terms of the legal definition of a standard, then, the 1898 law was wanting. Not until ranges of error and tolerance were specified in 1901 did electrical standards, in their fullest dimensions, exist in Germany.[135] The 1898 law had accommodated a different approach to precision, one based on consensus, but it was still the case that without error, there could be no legal standard.

CONCLUSIONS

The German settlement on the ohm was temporary and limited. The 1898 law was not immediately put into effect; it was later superceded. The German community of scientists and electrotechnologists embraced the settlement incompletely. Nevertheless, the settlement is a benchmark in the history of German precision measurement and standards. Earlier legal standards evolved from tangible experiences with the size, shape, and weight of objects in everyday life; the ohm derived from a scientific theory. In the first half of the century legal standards were based on the precise measurements of one scientist or at most a few often working in cooperation with instrument makers or mechanics. The German establishment of the ohm, by contrast, was a collective effort, but not necessarily a cooperative one. The German case of the legal ohm is thus revealing of both how agreements on refined measurements were achieved and how precision was redefined to meet special needs in a new kind of standards construction.

So well established by the 1860s were the foundations of and practices in German weights and measures that framers of new supplementary standards were, from a legal perspective, all but obliged to relate to these prior practices, if not to adopt them entirely. To a certain degree stages in the establishment of electrical standards were in fact *geprägt* – stamped – by weights and measures reform which drew upon longstanding economic, social, and legal traditions in the state's use of quantification. Precisely when German scientists and electricians were creating electrical units and standards, officials from several German states met to consider how the Germanies could establish a unified system of weights and measures to facilitate trade. This movement, initiated by a series of recommendations made in Frankfurt in 1860, culminated at the end of the decade with the adoption of the metric system by the North German Confederation. The recommendations made along the way demonstrated how deeply engrained were certain experimental practices and legal specifications in German weights and measures, and why, in particular, the German Reich was unlikely to adopt the English way of setting standards – either legally or scientifically.

Social meanings of standard determinations thus factored into the German–British exchange, revealing how reaching agreement across national boundaries was more than a matter of numbers. When the Germans and British argued with one another, they were debating not numerical values, but the means by which those numbers had been produced. What this study of local cultures demonstrates is in one sense obvious: that the adaptation of foreign practices, including in precision measurement, occurs in a manner that preserves the salient, and perhaps most revealing, features of a local culture, an observation made frequently enough by anthropologists. Transformations occasioned by investigations like Dorn's compel us to consider more closely not only the character of local scientific cultures, but also the manner in which they interact with one another. For the case of the determination of the German resistance standard, that interaction resulted in an altered understanding of precision which in turn created problems in the legal realm: in the definitions of crime and punishment and so in the understanding of the moral standards of normalcy and deviance. Modern standards do have a social meaning, one no less rich than that of their early modern predecessors.

Centre for German and European Studies, Georgetown University

NOTES

I am grateful for the assistance of Dr. Helmut Rohlfing of the Niedersächsische Staats- und Universitätsbibliothek and Dr. Regina Mahlke of the Staatsbibliothek zu Berlin-Preussischer Kulturbesitz, Haus Zwei. For their comments, I thank Larry Lagerstrom, Myles Jackson, Jed Buchwald and audiences at the Dibner Institute for the History of Science and Technology, Massachusetts Institute of Technology; the University of Oklahoma; and the History of Science Society. I would like to thank the National Science Foundation (DIR-9023476) and the National Endowment for the Humanities (RH-21005-91) for their support.

[1] Witold Kula, *Measures and Men* (Princeton: Princeton University Press, 1986), 3–4.
[2] Kula, *Measures and Men*, 118. Also see David Cahan, *An Institute for an Empire: The Physikalisch–Technische Reichsanstalt, 1871–1918* (Cambridge: Cambridge University Press, 1989), 5.
[3] On authority and quantification, see the excellent discussions in Theodore M. Porter, "Objectivity and Authority: How French Engineers Reduced Public Utility to Numbers", *Poetics*

Today 12 (1991):245–65; idem, *Trust in Numbers: The Pursuit of Objectivity in Science and Public Life* (Princeton: Princeton University Press, 1995).

[4] Kathryn M. Olesko, "The Meaning of Precision: The Exact Sensibility in Early Nineteenth Century Germany", in *The Values of Precision*, ed. M. Norton Wise (Princeton: Princeton University Press, 1995), 105–34.

[5] "Entwurf eines Gesetzes, betreffend die elektrischen Masseinheiten", *Stenographische Berichte über die Verhandlungen des Reichstages*, Aktenstuck Nr. 181, IX. Legislatur–periode, V. Session, 1897/98, Bd. 164, 1735–41, on 1736.

[6] The British case is discussed by: Bruce Hunt, "The Ohm is Where the Art Is: British Telegraph Engineers and the Development of Electrical Standards", *Osiris* 9 (1994):48–63; Larry Randles Lagerstrom, "Universalizing Units: The Rise and Fall of the International Electrical Congress, 1881–1904", unpubl. ms., 1992, 3–5; Simon Schaffer, "Late Victorian Metrology and its Instrumentation: A Manufactory of Ohms", in *Invisible Connections: Instruments, Institutions, and Science*, ed. Robert Bud and Susan E. Cozzens (Bellingtam, Wa.: SPIE Optical Engineering Press, 1992), 23–56; Crosbie Smith and M. Norton Wise, *Energy and Empire: A Biographical Study of Lord Kelvin* (Cambridge: Cambridge Univ. Press, 1989), 684–98; Hunt, "'Practice vs. Theory': The British Electrical Debate, 1888–1891", *Isis* 74 (1983):341–55; Schaffer, "'Accurate Measurement is an English Science'", in *The Values of Precision*, 137–72; Graeme Gooday, "Precision Measurement and the Genesis of Physics Teaching Laboratories in Victorian Britain", *British Journal for the History of Science* 23 (1990):25–51; idem, "Teaching Telegraphy and Electrotechnics in a Physics Laboratory: William Ayrton and the Creation of an Academic Space for Electrical Engineering in Britain, 1873–1884", *History of Technology* 13 (1991):73–111. In the preparation of this essay, I was especially influenced by Bruce Hunt's recent work.

[7] Wilhelm Weber, *Elektrodynamische Maassbestimmungen: Ueber ein allgemeines Grundgesetz der elektrischen Wirkung* (1846), in *Wilhelm Weber's Werke*, ed. by Kgl. Gesellschaft der Wissenschaften zu Göttingen, 6 vols. (Berlin: J. Springer, 1892–1894), 3:25–214. On the transformation in German physics, see Kenneth L. Caneva, "From Galvanism to Electrodynamics: The Transformation of German Physics and its Social Context", *Historical Studies in the Physical Sciences* 9 (1978):63–159.

[8] Rudolf Stichweh, *Zur Entstehung des modernen Systems wissenschaftlicher Disziplinen: Physik in Deutschland, 1740–1890* (Frankfurt a.M.: Suhrkamp, 1984), 203–5, 231–2, 379–81, 480–2; Christa Jungnickel and Russell McCormmach, *Intellectual Mastery of Nature: Theoretical Physics from Ohm to Einstein*, 2 vols. (Chicago: Univ. of Chicago Press, 1986), 1:75–7, 139–49; Kathryn M. Olesko, *Physics as a Calling: Discipline and Practice in the Königsberg Seminar for Physics* (Ithaca, N.Y.: Cornell University Press, 1991), 173–4, 185–6, 188, 279, 409–10, 455; idem, "Tacit Knowledge and School Formation", *Osiris* 8 (1993):15–29.

[9] *Weber's Werke*, 2:20–42, 3:31, passim.

[10] Moritz Jacobi, quoted in *Weber's Werke*, 3:303–4.

[11] Oskar Frölich, *Die Entwickelung der elektrischen Messungen*. Die Wissenschaft: Sammlung naturwissenschaftlicher und mathematischer Monographien, Heft 5 (Braunschweig: F. Vieweg & Sohn, 1905), 96.

[12] Wilhelm Weber, "Messungen galvanischer Leitungswiderstände nach einem absolute Maasse", *Annalen der Physik und Chemie* [hereafter *AP*] 8 (1851):337–69, on 357.

[13] Weber, "Messungen galvanischer Leitungswiderstände", 337, 358.

[14] Olesko, "The Meaning of Precision", 118. Weber's interest in rectifying standards of weight is also evident in his work on the specific gravity of water. Finding that difference in various determinations of weight were too great to be attributed to errors of observation or calculational errors, Weber suggested that uncertainty in the specific gravity of water was the cause of discrepancies in calculations used to determine standards of weight. Weber, "Ueber die noch vorhandene Unzuverlässigkeit im specifischen Gewichte des Wassers" (1830), *Weber's Werke*, 1:416–18.

[15] On Gauss, see Olesko, "The Meaning of Precision", 119–21.

[16] Weber, "Electrodynamische Maassbestimmungen insbesondere Widerstandsmessungen" (1852), *Weber's Werke*, 3:301–471, on 303.

[17] Weber, "Elektrodynamische Maassbestimmungen", *Weber's Werke*, 3:350–4.

[18] Weber, "Messungen galvanischer Leitungswiderstände", 357.

[19] Weber, "Elektrodynamische Maassbestimmungen", *Weber's Werke*, 3:318.

[20] Frölich, *Die Entwickelung der elektrischen Messungen*, 153, 43, 96.

[21] See esp. *Weber's Werke*, 2:20–42, 274–6.

[22] Weber, "Elektrodynamische Maassbestimmungen", *Weber's Werke*, 3:355. Weber's effort to streamline experimentation and reduce the time-investment in routine tasks, such as instrument

reading, is evident in his many studies on instruments. See for instance, Weber, "Ueber Barometer–und Thermometerskalen" (1837), *Weber's Werke*, 1:516–25, esp. 516.

[23] William Thomson, "Ueber die Elektrizitäts–Leitungsfähigkeit von käuflichen Kupferdräthen aus verschiedenen Bezugsquellen", *Zeitschrift des deutsch–österreichischen Telegraphen–Vereins* [hereafter *ZTV*] 5 (1858):137–41, on 140.

[24] E. Lampe, "Philipp Wilhelm Brix", *Verhandlungen der Deutschen Physikalischen Gesellschaft* 1 (1899):125–35, esp. 130–2 (quote, 131); Olesko, Physics as a Calling, 137–9.

[25] C. Frischen, "Ueber Isolation oberirdischer Leitungen", *ZTV* 5 (1858):51–3; "Nachbemerkung der Redaction", ibid.:53–4, on 54.

[26] Werner Siemens, "On the Question of the Unit of Electrical Resistance", *Philosophical Magazine* 31 (1866):325–36.

[27] Wilhelm Beetz, "Das Ohm'sche Gesetz mit Beispielen seiner Anwendung in der Telegraphie", *ZTV* 2 (1855):49–58, 73–81, on 54.

[28] Christoph Bergeat, "Ueber die Bestimmung der Factoren eines galvanischen Stromes und einen hierzu sehr bequemen Rheostaten", *ZTV* 4 (1857):265–75.

[29] Frölich, *Die Entwickelung der elektrischen Messungen*, 100–1.

[30] Carl Siemens, "Bestimmung der auf Telegraphenleitungen vorkommenden Störungen mittelst Differential–Instrumenten von Siemens und Halske", *ZTV* 5 (1858):13–18, on 13, 14, 17.

[31] Ernst Esselbach, "Notiz über den Leitungswiderstand einiger Unterseekabeln", *ZTV* 6 (1859):109–10. Esselbach did not include a temperature specification.

[32] Werner Siemens, "Vorschlag eines reproducibaren Widerstandsmasses", *ZTV* 7 (1860):55–68. On Siemens's unit, see Frölich, *Die Entwickelung der elektrischen Messungen*, 96–7; Lagerstrom, "Universalizing Units", 2–3.

[33] Siemens, "Vorschlag", 65, 59, 61, 63.

[34] Siemens, "Vorschlag", 56, 64, 65.

[35] Werner Siemens, "Widerstands–Etalon", *AP* 120 (1863):512. Siemens had issued graduated resistance coils, in units from 1 to 100, as early as 1848; he did not specify the foundation of the unit. Siemens, "On the Question of the Unit of Electrical Resistance", 336.

[36] Werner Siemens, "Proposal for a New Reproducible Standard Measure of Resistance to Galvanic Circuits", *Philosophical Magazine* 21 (1861):25–38.

[37] Esselbach, Kirchhoff, and Siemens argued their points in separate letters to the Committee on Electrical Standards in 1862, rpt. in *Reports of the Committee on Electrical Standards*, ed. Fleeming Jenkin (London: E. & F. N. Spoon, 1873), 26–32.

[38] "First Report – Cambridge, October 3, 1862", rpt. in *Reports*, 1–11, on 2, 5, 7. On the establishment of the British resistance standard, see also Hunt, "The Ohm is Where the Art Is"; Schaffer, "Late Victorian Metrology"; Smith & Wise, *Energy and Empire*, 684–98.

[39] See for instance, Augustus Matthiessen, "Ueber die elektrische Leitungsfähigkeit der Metall–Legirungen", *ZTV* 7 (1860):9–14; idem, "Ueber eine Legirung, welche als Widerstandsmaass gebraucht werden kann", ibid., 73–5; A. Matthiessen and M. Holzmann, "Ueber die Leitungsfähigkeit des reinen Kupfers und deren Verminderung durch Metalloide und Metalle", *ZTV* 7 (1860):261–9; W. Beetz, "Ueber die Elektricitätsleitung durch Kohle und durch Metalloxyde", *ZTV* 7 (1860):270–1.

[40] "Second Report – Newcastle-on-Tyne, August 26, 1863", rpt. in *Reports*, 39–53, on 53. Augustus Matthiessen and Charles Hockin, "On the Construction of Copies of the B.A. Unit", rpt. in *Reports*, 135–7, on 136.

[41] Siemens, "On the Question of the Unit of Electrical Resistance", 326; Werner Siemens, "Ueber Widerstandsmaasse und die Abhängigkeit des Leitungswiderstandes der Metalle von der Wärme", *ZTV* 8 (1861):76–85, on 77. In consequence of his approach, Siemens collapsed the unit and the standard into one, while the British kept them separate. Lagerstrom, "Universalizing Units", 4 (fn.12). Before the establishment of the meter in France at the end of the eighteenth century and scientific standards in the Germanies in the first half of the nineteenth, however, standards and units were rarely separated legally.

[42] Matthiessen later found German silver acceptable as a distributable standard, but still refused to use mercury for the original calibration. Augustus Matthiessen, "Einige Bemerkungen zu der Abhandlung des Hrn. Siemens: Ueber Widerstandsmaasse und die Abhängigkeit des Leitungswiderstandes der Metalle von der Wärme", *AP* 114 (1861):310–21.

[43] Augustus Matthiessen, "On the Specific Resistance of the Metals in terms of the B.A. Unit (1864) of Electric Resistance, together with some Remarks on the so-called Mercury Unit", *Philosophical Magazine* 29 (1865):361–70, on 367–8.

[44] "First Report", rpt. in *Reports*, 5 (emphasis added).

[45] That the British were indeed thinking in these terms is also evident in their remark that they hoped copies of the proposed material standard "will soon be everywhere obtainable, and that a man will no more think of producing his own standard than of deducing his foot-rule from a pendulum.."... "First Report", rpt. in *Reports*, 5.

[46] Siemens, "Ueber Widerstandmaasse und die Abhängigkeit des Leitungswiderstandes der Metalle von der Wärme", 76.

[47] Fleeming Jenkin, "New Unit of Electrical Resistance", *Philosophical Magazine* 29 (1865):477–86, on 484.

[48] Olesko, "The Meaning of Precision".

[49] On this culture of precision, see Kathryn M. Olesko, "The Culture of Precision in Nineteenth Century Germany", unpubl. paper presented at the Deutsches Historisches Institut, Washington, DC, 21 May 1992; idem, "The Meaning of Precision"; idem, "Precision, the Cadaster, and Property Rights in Pre–Imperial Germany", essay presented at the combined meeting of the British Society for the History of Science and the History of Science Society, Edinburgh, July 1996. Simon Schaffer has detailed attributes of British precision measurement in "Accurate Measurement is an English Science." A discussion of American precision pertinent to present concerns is found in Francis B. Crocker, "The Precision of Electrical Engineering", *Transactions of the American Institute of Electrical Engineers* 14 (1897):237–49. On concepts of precision in early modern Europe, see Steven Shapin, *A Social History of Truth: Civility and Science in Seventeenth Century England* (Chicago: University of Chicago Press, 1994), 310–54; and on the modern culture of precision, see the articles in *The Values of Precision*, ed. M. Norton Wise (Princeton: Princeton University Press, 1995).

[50] "Second Report", rpt. in *Reports*, 51; "Third Report – Bath, September 14, 1864", rpt. in *Reports*, 110–14, on 111; Siemens, "On the Question of the Unit of Electrical Resistance", 334.

[51] Ian Hopley, "Maxwell's Work on Electrical Resistance: the Determination of the Absolute Unit of Resistance", *Annals of Science* 13 (1957):265–72; Schaffer, "Accurate Measurement is an English Science".

[52] "Second Report", rpt. in *Reports*, 47.

[53] J. Clerk Maxwell, Balfour Stewart, and Fleeming Jenkin, "Description of an Experimental Measurement of Electrical Resistance, made at King's College" [1863], rpt. in *Reports*, 96–109, on 96 (emphasis added); see also "First Report", ibid., 7.

[54] "Second Report", rpt. in *Reports*, 50.

[55] Maxwell et al., "Description of an Experimental Measurement", 96–109.

[56] "Second Report", rpt. in *Reports*, 40.

[57] "Second Report", rpt. in *Reports*, 50–1.

[58] "Third Report", rpt. in *Reports*, 111.

[59] "Third Report", rpt. in *Reports*, 111.

[60] Siemens, "On the Question of the Unit of Electrical Resistance", 327.

[61] Siemens, "On the Question of the Unit of Electrical Resistance", 327, 329.

[62] Maxwell et al., "Description of an Experimental Measurement", 105, 108, 109.

[63] Matthiessen, "On the Specific Resistance of the Metals".

[64] Siemens, "On the Question of the Unit of Electrical Resistance", 328–9 (emphasis added).

[65] Augustus Matthiessen, "Note on Dr. Siemens's paper 'On the Question of the Unit of Electrical Resistance' ", *Philosophical Magazine* 31 (1866):376–80.

[66] Fleeming Jenkin, "Reply to Dr. Werner Siemens's Paper 'On the Question of the Unit of Electrical Resistance' ", *Philosophical Magazine* 32 (1866):161–77.

[67] Jenkin, "Reply", 176–7 (emphasis added).

[68] Johann Eytelwein, *Vergleichungen der gegenwärtig und vormals in den königlichen Preussischen Staaten eingeführten Maasse und Gewichte, mit Rücksicht auf die vorzüglichsten Maasse und Gewichte in Europa* (Berlin: Realschulbuchhandlung, 1810^2).

[69] *Gutachten über Einführung gleichen Masses und Gewichtes in den deutschen Bundesstaaten. Ausgearbeitet von der durch die hohe deutsche Bundesversammlung hierzu berufenen Kommission* (Frankfurt am Main: Bundes–Druckerei, 1861).

[70] *Gutachten*, 10, 66–8.

[71] *Gutachten*, 15.

[72] Jenkin, "Reply", 163.

[73] *Gutachten*, 40–41. See also Gustav Karsten, *Die international General–Konferenz für Maass und Gewicht in Paris 1889: Rede gehalten beim Antritt des Rektorates der Universität Kiel am 5 März 1890* (Kiel: Universitäts–Buchhandlung, 1890), 6. Karsten and others point out that the exact and compelling nature of Bessel's work foreclosed, in Prussia, consideration of alternative standards of measure, including the meter, whose foundations were considered less exactly established.

[74] Friedrich Wilhelm Bessel, *Untersuchungen über die Länge des einfachen Secundenpendels, besonders abgedruckt aus den Abhandlungen der Akademie zu Berlin für 1826* (Berlin: Kgl. Akademie der Wissenschaften, 1828). At Königsberg, Bessel's methods, as exemplified in the seconds pendulum investigation, became the foundation of Franz Neumann's teaching program in theoretical physics in which the analysis of instruments, error analysis, and the method of least squares were central. Olesko, *Physics as a Calling*, esp. chp. 4: Mechanics and the Besselian Experiment.

[75] Friedrich Wilhelm Bessel, *Darstellung der Untersuchungen und Maassregeln, welche, in den Jahren 1835 bis 1838, durch die Einheit des Preussischen Längenmaasses worden sind* (Berlin: Kgl. Akademie der Wissenschaften, 1839); Olesko, "The Meaning of Precision", 121–5.

[76] *Gutachten*, 40–3, 64–8.

[77] *Gutachten*, 53.

[78] *Gutachten*, 53–4.

[79] *Gutachten*, 54–5.

[80] Gustav Karsten, *Ueber die Maass- und Gewichts-Ordnung für den Norddeutschen Bund* (Kiel: A. F. Jensen, 1869).

[81] *Wörterbuch des Deutschen Staats- und Verwaltungsrechts*, ed. Max Fleischman, 3 vols. (Tübingen: J. C. B. Mohr, 1911–1914), 2:813–18 (s.v. "Mass und Gewicht"), on 814–15.

[82] Weber, "Ueber die beabsichtigte Einführung eines galvanischen Widerstands–Etalons oder Standards" (1861), *Weber's Werke*, 4:10–16, on 13.

[83] Weber, "Ueber einheitliche Maasssysteme" (1861), *Weber's Werke*, 1:526–39, on 534 (emphasis in original).

[84] Weber, "Zur Galvanometrie" (1862), in *Weber's Werke*, 4:17–96, on 69.

[85] Louis Schwendler [electrician at Siemens Brothers in Woolwich], "Ueber den passendsten Widerstand des bei Messungen mit der Wheatstone'schen Brücke benutzten Galvanometers", *ZTV* 13 (1866):77–82, 14 (1867):32–8; Franz Dehms [Siemens & Halske, Berlin], "Vorschlag zu einer veränderten Construction der Wheatstone'schen Brücke, und Bemerkungen über die Messung mit derselben", *ZTV* 13 (1866):259–70; C. W. Siemens, "Ueber einen Widerstandmesser", *ZTV* 14 (1867):76–8; Franz Dehms [now with the Prussian state telegraph bureaucracy in Berlin], "Methode zur Herstellung von Widerstandsscalen sowie Bemerkungen über Anordnung derselben", *ZTV* 14 (1867):4–14; Werner Siemens, "Das Universal–Galvanometer", *ZTV* 15 (1868):1–6; Heinrich Weber [Technische Hochschule, Braunschweig], "Vorschriften zur Construction von Galvanoskopen, welche das Maximum der Empfindlichkeit besitzen", *ZTV* 16 (1869):105–14. I discuss changes brought to instrumentation and exact experiment as a result of technical demands in "Industrial Demands and the Political Economy of Exact Experiment", unpubl. essay presented at the Conference on Writing the History of Physics, St. John's College, Cambridge, 3–5 April 1991.

[86] During the 1860s the practical problem in telegraphy most frequently discussed was fault location. See *ZTV*, volumes 7–16 (1860–69).

[87] See e.g., C. Siemens, "Ueber einen Widerstandsmesser", 76.

[88] In his redesigning of resistance scales, for instance, Franz Dehms tried make available to electricians the accuracy that physicists had in their system of weights. Dehms, "Method zur Herstellung von Widerstandsscalen", 4.

[89] W. Siemens, "Das Universal-Galvanometer", 1.

[90] Franz Dehms, "Neue Bestimmung der Siemens'schen Widerstands–Einheit", *ZTV* 15 (1868):13–44, on 13, 33. The *ZTV* publication of his investigation became Dehms's doctoral dissertation at Rostock; Franz Dehms, "Ueber eine Reproduction der Siemen'schen Widerstands–Einheit", *Inaugural–Dissertation zur Erlangung der philosophischen Doctorwürde bei der Universität zu Rostock* (Berlin: Ernst & Korn, 1868). Note that in his doctoral dissertation he drew attention to the reproducibility of the Siemens unit, one element in the British–German controversy; the text remained the same.

[91] "Third Report", rpt. in *Reports*, chart facing 114.

[92] Dehms, "Neue Bestimmung", 40.

[93] Wilhelm Weber Nachlass, Nrs. 21.1–21.5: Seminar-Vorlesungen in Nachschrift v. K. Hattendorff; Hermann Wagner Nachlass, Nrs. 6.1–6.5: Vorträge von Wilhelm Weber über verschiedene Gegenstände der mathematischen Physik, gehalten im physikalischen Seminar der Georgia Augusta, 1860–63; both located in the Abteilung für Handschriften und seltene Drucke, Niedersachsische Staats- und Universitätsbibliothek, Altbau, Göttingen. Electrical measurements constituted the core of Friedrich Kohlrausch's pedagogic physics, which was based on the Göttingen program in physics instruction. Friedrich Kohlrausch, *Leitfaden der praktischen Physik zunächst für das physikalische Prakticum in Göttingen* (Leipzig: B.G. Teubner, 1870), esp. 60–91; Olesko, "Tacit Knowledge and School Formation", 26–8. Copies of the electrical resistance units of

Siemens and the British Association can still be found in the I. Physikalisches Institut, Universität Göttingen.

94 Friedrich Kohlrausch, "Beobachtungen im magnetischen Observatorium aus dem Jahre 1869, insbesondere Bestimmung der Siemens'schen Widerstandseinheit nach absolutem Masse" (1870), in *Gesammelte Abhandlungen von Friedrich Kohlrausch*, ed. W. Hallwachs, Adolf Heydweiller, K. Strecker, and O. Wiener, 2 vols. (Leipzig: J. A. Barth, 1910–11), 1:383–91, on 383.

95 Friedrich Kohlrausch, "Determination of the Absolute Value of the Siemens's Mercury Unit of Electrical Resistance", *Philosophical Magazine* 47 (1874):294–309, 342–54, on 294. I have made some adjustments in the English translation based on a comparison with the original German; cf. Kohlrausch, *Gesammelte Abhandlungen*, 1:435–68, on 435.

96 Kohlrausch, "Determination", 295 [1:436], 297, 301, 299 [1:440], 299, 300. Among the possible sources of the British data's irregularity, Kohlrausch cited problems with: the rotational velocity of the needle; the inductor's motion to the left and right; the thread's torsion; elastic yielding; the metal parts of the apparatus; and others. The principal "share of the uncertainty" was in his view due to the horizontal component of the earth's magnetic force. Ibid., 305.

97 Kohlrausch, "Determination", 309.

98 Kohlrausch, "Determination", 353.

99 Henry Rowland, "Note on Kohlrausch's Determination of the Absolute Value of the Siemens Mercury Unit of Electrical Resistance", *Philosophical Magazine* 50 (1875):161–3.

100 Oskar Frölich, *Die Lehre von der Elektricität und dem Magnetismus mit besonderer Berücksichtigung ihrer Beziehungen zur Telegraphie. Handbuch der elektrischen Telegraphie*, ed. K. E. Zetzsche, vol. 2 (Berlin: J. Springer, 1878²), 64, 98, 96, 442–3, 437; on 437, 443.

101 These investigations include: Werner Siemens, "Directe Messung des Widerstandes galvanischer Ketten", *AP Jubelband* (1874):445–8; Oskar Frölich, "Directe Messung des Widerstandes galvanischer Ketten [nach Siemens]", ibid.:448–52; F. Himstedt, "Ueber eine Methode zur Bestimmung des Ohm", *AP* 22 (1884):281–6; Heinrich Wild, "Bestimmung des Werthes der Siemens'schen Widerstands–Einheit in absolutem elektromagnetischen Maasse", *Mémoires l'Académie impériale des sciences de St. Petersbourg* 32 (1884):1–122; H. Wild, "Bestimmung des Werthes der Siemens'schen Widerstandseinheit in absolutem elektromagnetischen Maasse", *AP* 23 (1884):665–77; F. Kohlrausch, "Zu einigen Kritischen Bemerkungen des Hrn. Wild", *AP* 23 (1884):344–8; H. Wild, "Antwort auf einige Bemerkungen des Herrn F. Kohlrausch", *AP* 24 (1885): 209–14; Stefan Lindeck, "Ueber eine Herstellung von Normalquecksilberwiderstände", *Zeitschrift für Instrumentenkunde* 11 (1891): 171–83. Other investigations are cited below.

102 Gustav Wiedemann, "Ueber die bisherigen Methoden zur Feststellung des Ohm", *Elektrotechnische Zeitschrift* [hereafter *EZ*] 3 (1882):260–9, on 260, 266.

103 Siemens & Halske, "Bericht über eine Rekonstruktion der Quecksilverwiderstandseinheit und Beschreibung der Einrichtungen für elektrische Messungen", *EZ* 3 (1882):408–15, on 408.

104 In the Siemens & Halske reconstruction of the mercury unit, for instance, it was found that the "greatest uncertainty" arose in the determination of the weight of the mercury, but the bureau of weights and measures was able to determine that the uncertainty was only 1/10,000 of the final value. Siemens & Halske, "Bericht", 410.

105 Siemens & Halske, "Bericht".

106 Wilhelm Weber and Friedrich Zöllner, "Ueber Einrichtungen zum Gebrauch absoluter Maasse in der Elektrodynamik mit praktischer Anwendung" (1880), in *Weber's Werke*, 4:420–76, esp. 431, 436, 439.

107 Wiedemann, "Ueber die bisherigen Methoden zur Feststellung des Ohm", 269, 268.

108 Wiedemann, "Ueber die bisherigen Methoden zur Feststellung des Ohm", 268.

109 Siemens & Halske, "Bericht", 408.

110 Lord Rayleigh, "Address to the Mathematical and Physical Sciences Section", *British Association Reports* (1882): 437–41, on 438.

111 Schaffer, "Accurate Measurement is an English Science", 164. Schaffer emphasized the collaborative nature of British precision measurement in a handout of quotations, with generalizations, that accompanied his initial presentation of his essay at the Princeton Workshop in History of Science on "Values of Precision" in March 1992.

112 J. Clerk Maxwell and Fleeming Jenkin, "On the Elementary Relations between Electrical Measurements", *Philosophical Magazine* 29 (1865):436–60, 507–25, on 437.

113 These differences help in interpreting Henry Rowland's 1876 remark, that "the accurate measurement of resistance either absolutely or relatively is an English science almost unknown in Germany" (Quoted in Schaffer, "Accurate Measurement is an English Science,", 139. Schaffer draws his theme from this remark.) For Rowland, the activity of "accurate measurement" appears to have meant something different than it did in Germany. Rowland's remark was accompanied by

the observation that, with regard to the apparatus used at Göttingen, "the form of the earth inductor was such that it would be impossible to find its area with accuracy" (this part of Rowland's comment was eliminated from the final published version of Schaffer's essay). Rowland thus appears to have recoiled from the prospect of a complicated process of error analysis, a conclusion supported by Wiedemann's observations that not only did Rowland analyze his data improperly, but he also failed to take into account several sources of constant errors in his experiment. Wiedemann, "Ueber die bisherigen Methoden zur Feststellung des Ohm", 266.

[114] David Cahan, "Werner Siemens and the Origin of the Physikalisch-Technische Reichsanstalt, 1872–1887", *Historical Studies in the Physical Sciences* 12 (1982):253–83, on 259.

[115] Lagerstrom, "Universalizing Units", 8. Lagerstrom concludes that the main factors standing in the way of agreement were: (1) the persistence of nationalist sentiments, especially concerning the sovereign right of a state to control its own standards, in the context of trying to establish *international* units; and (2) the format and lack of authority of the Congresses.

[116] Ernst Dorn, "Die elektrotechnischen Versuche auf der internationalen Elektrizitäts–Ausstellung in München, 1882", *EZ* 4 (1883):404–15, on 404.

[117] Die Königliche Akademie der Wissenschaften [Curtius, Emil du Bois–Reymond, Mommsen, Auwers] to Minister von Gossler, 12 February 1885, Geheimes Staatsarchiv Preussischer Kulturbesitz [hereafter GSPK], Berlin–Dahlem, Ministerium der geistlichen, Unterrichts– u. Medicinal–Angelegenheiten. Unterrichts–Abteilung. Acta betreffend: Die internationale Conferenz zur Berathung von Fragen der elektrischen Wissenschaft und Praxis sowie die elektrische Ausstellung in Paris, desgl. in München, Rep. 76vc, Sekt 1, Tit. 11, Teil VI, Nr. 5, Bd. II, 1883–91, fols. 271–274, on fol. 273r; "Electrische Einheiten und Lichteinheiten", *AP* 22 (1884): 616.

[118] Akademie der Wissenschaften to Gossler, 12 February 1885; Werner Siemens, "Ueber elektrische und Lichteinheiten nach den Beschlüssen der Pariser internationalen Conferenz" (1884)", in Werner Siemens, *Wissenschaftliche und Technische Arbeiten*. Erster Band. *Wissenschaftliche Abhandlungen und Vorträge* (Berlin: J. Springer, 1889), 399–403, on 401.

[119] Lagerstrom, "Universalizing Units", 13–20.

[120] Die Königliche Akademie der Wissenschaften [Auwers, Emil du Bois–Reymond, Curtius, Mommsen] to Kultusministerium, 6 April 1883, GSPK, Internationale Conferenz, fols. 114–16, on fol. 114–14r, 116 (emphasis added).

[121] Akademie der Wissenschaften to Minister von Gossler, 12 February 1885, fols. 273–273r.

[122] Cahan, *An Institute for an Empire.*

[123] Frölich, *Die Entwickelung der elektrischen Messungen*, 98. The Physikalisch–Technische Reichsanstalt's contribution to electrical measurement, standards, and testing, was extensive; see Cahan, *An Institute for an Empire*, passim.

[124] Cahan, *An Institute for an Empire*, 43, 104–105, 109, 115; Johann Pernet, "Ueber die physikalisch-technische Reichsanstalt zu Charlottenburg und die daselbst ausgeführten electrischen Arbeiten", *Schweizerische Bauzeitung* 18 (1891):1–6, esp. 4; idem, "Ueber den Einfluss physikalischer Präcisionsmessungen auf die Förderung der Technik und des Mass– und Gewichtswesens", ibid., 24 (1894):110–14, esp. 112.

[125] Cahan, *An Institute for an Empire*, 81.

[126] "Vorschläge zu gesetzlichen Bestimmungen über elektrische Maasseinheiten entworfen durch das Curatorium der Physikalisch-Technischen Reischsanstanlt", in Friedrich Ernst Dorn, *Vorschläge zu gesetzlichen Bestimmungen über elektrische Maasseinheiten, entworfen durch das Curatorium der Physikalisch-Technischen Reichsanstalt: Nebst kritischem Bericht über den wahrscheinlichen Werth des Ohm nach dem bisherigen Messungen* (Berlin: J. Springer, 1893), 12.

[127] Olesko, *Physics as a Calling*, 349, 350–2, 357; Albert Wigend, "Ernst Dorn", *Physikalische Zeitschrift* 17 (1916):297–9, esp. 298 (work on ohm).

[128] Ernst Dorn, "Die Reduction der Siemens'schen Einheit auf absolutes Maass", *AP* 17 (1882):773–816, on 773, 774, 785.

[129] Ernst Dorn, "Eine Bestimmung des Ohm", *AP* 36 (1889):22–72, 398–446.

[130] Ernst Dorn, "Ueber den wahrscheinlichen Werth des Ohm nach den bisherigen Messungen", in Dorn, *Vorschläge*, 19–86, on 84, 86.

[131] "Vorschläge zu gesetzlichen Bestimmungen über elektrische Maasseinheiten", in Dorn, *Vorschläge*, 14. This error made the actual value somewhat higher than Dorn's, and so the entire spread of values shifted, making 1.063 the "best" choice.

[132] "Entwurf eines Gesetzes betreffend die elektrischen Masseinheiten", *Stenographische Berichte über die Verhandlungen des Reichstages*, Aktenstuck Nr. 181, IX. Legislatur–periode, V. Session, 1897/98, Bd. 164, 1735–41, on 1737 (emphasis added). This point had been made in the Physikalisch-Technische Reichanstalt's 1893 recommendation. "Vorschläge zu gesetzlichen Bestimmungen über elektrische Maasseinheiten", in Dorn, *Vorschläge*, 7–18, on 8. Dorn emphasized that

German delegates to the 1881 International Electrical Congress in Paris had stressed this issue as well.
[133] "Entwurf", 1738.
[134] "Entwurf", 1736, 1740, 1741. The problem of defining crime in the marketplace of electricity was not peculiar to Germany. In 1897 the American electrical engineer Francis B. Crocker reported: "It has also been held by many courts, that electricity being intangible, [it] has no real existence, so that tapping of current from wires could hardly be considered as a theft, except in an imaginary sort of way. The production of electrical energy in central stations has been decided by metaphysically inclined judges to be a totally different kind of business from the manufacture of gas". Crocker, "The Precision of Electrical Engineering", 238.
[135] *Wörterbuch des Deutschen Staats- und Verwaltungsrechts*, 2:817.

SYNTONY AND CREDIBILITY:
JOHN AMBROSE FLEMING, GUGLIELMO MARCONI, AND
THE MASKELYNE AFFAIR

The witnessing and reporting of successful experiments held a special strategic place in Guglielmo Marconi's early wireless telegraphy, for they advertised the practical nature of his system and calmed objections to it. When, after 1900, syntony, or tuning, became a central issue in radiotelegraphy, Marconi devised a new syntonic system for sending and receiving wireless messages, and provided a series of powerful demonstrations of the system's effectiveness. These were witnessed and then reported by John Ambrose Fleming, scientific advisor to the Marconi Company since 1899. Fleming was able to act as a trustworthy witness because he had high credibility in the British electrical engineering and physics communities – a credibility that was built upon twenty years during which he acted as a mediator between the worlds of alternating-current power engineering and of physics. Fleming, as a supportive witness, was thus troublesome to Marconi's adversaries. In June 1903, Nevil Maskelyne, one of Marconi's opponents, interfered with Fleming's public demonstration of Marconi's syntonic system at the Royal Institution by sending derogatory messages from his own simple transmitter. This incident, which became known as the Maskelyne affair, severely damaged both Marconi's and Fleming's credibility. Indeed, soon after the affair Fleming was dismissed from his advisorship to Marconi.

The Maskelyne affair has been mentioned by several historians, but it has never been fully placed in all of its relevant contexts. By contexts I mean three associated circumstances – the scientific and technological context in which early syntonic devices were understood and developed; the corporate context in which the competition for monopolizing the market for wireless telegraphy became intense; and finally the authorial context in which issues concerning who should hold authority in the rapidly developing field of wireless telegraphy became prominent. Detailed examination of the Maskelyne affair within these contexts will uncover several issues worth exploring for the history of early wireless telegraphy: these include the struggle between Marconi and his opponents; the efficacy of early syntonic devices; Fleming's role as a public witness to Marconi's private experiments; and the nature of Marconi's "shows". In addition, the affair provides a unique case study concerning the manner in which the credibility of engineers was created, consumed, and suddenly destroyed.

SYNTONY AND MARCONI'S "FOUR-SEVEN" (7777) PATENT

The theoretical basis of wireless telegraphy lay in the electromagnetic wave, which was first produced in the laboratory by Heinrich Hertz in 1888. To

J.Z. Buchwald (ed.), Scientific Credibility and Technical Standards in 19th and early 20th Century Germany and Britain, 157–176.

generate his waves Hertz, having charged a pair of condenser plates to high voltage, then discharged them through a spark across a small gap. This simple manipulation produced extremely high-frequency electromagnetic waves that propagated through space [Aitken 1976; Buchwald 1994]. Hertz had also used a spark-based device to detect his waves, but in the early 1890s more delicate detectors, such as coherers,[1] were devised. Few people, however, thought to utilize these devices for communication to produce, in effect, telegraphy without wires. The two major reasons for this neglect were: first, the maximum distance traversed by the wave, half a mile or so, was too short for any practical communication; and second, Hertzian devices seemed to be utterly dissimilar to telegraphic apparatus. Instead, scientists and engineers of the day found instrumental and conceptual similarities between the Hertzian device and light signalling apparatus used to communicate with ships.[2]

This changed in 1896. In that year Marconi increased the distance over which waves could be successfully detected up to several miles, and he also succeeded in sending messages using the telegrapher's Morse code. Though Marconi's success required several notable innovations, such as an increase in the power of the transmitter, improved sensitivity for the detector, and the use of a Morse telegraphic printer to record his transmitted signals, the most important one was the use of vertical antennas for the transmitter and the receiver. By increasing the height of his vertical antennae, Marconi could send messages farther and farther. Within a few years, he had sent a message over a hundred miles to sea, and this in itself sufficed to make wireless telegraphy a practical system for ship-to-shore communication.

Marconi's system using vertical antennas had, however, one serious disadvantage. It was powerful, but it was bad for syntony, or tuning. In practice, bad syntony meant that anybody who wanted to could easily capture messages by means of a simple detector. In theory, bad syntony meant that the wave generated from Marconi's transmitter did not have a narrowly-defined frequency range, but rather had a broad range of frequencies. According to Hertz's theory and experiment, the frequency (f) of the wave generated from a condenser–discharge circuit was (very nearly at least) determined only by its inductance (L) and capacitance (C), that is, frequency $f = 1/2\sqrt{(LC)}$. The wave generated from Marconi's vertical antenna transmitter should thus have had a well-defined frequency determined by its own inductance and capacitance, but it seemed not to be so.

A consensus emerged that the major factor at fault here lay in the fact that the wave generated by Hertz's or Marconi's devices was highly damped. Although opinions concerning the reason for the effect of damping at first varied, it was eventually decided that a highly damped electromagnetic wave could be thought of physically as well as mathematically as a superposition of many different waves, in fact an uncountably infinite sequence of them, each with its own frequency. Indeed, according to the mathematics of Fourier analysis, which was here granted direct physical significance, the more rapidly the wave was damped, the broader was its frequency range. This, the phenomenon termed "multiple resonance," was in fact regarded as one of the most difficult problems for Hertzian physics.[3] The more homogeneous the wave (or "continuous" as it was described by engineers of the day), the narrower its frequency range. Therefore, in order to secure good syntony one had to produce, or come close to producing, homogeneous or continuous waves.

Although it was not possible to produce continuous waves by means of a spark transmitter, there was a way to lessen damping. From the early 1890s, physicists had known that a transmitter in the shape of a loop was a more persistent vibrator than was a vertical line. In early 1897, while communicating with Silvanus P. Thompson on Marconi's wireless telegraphy, Lodge was reminded of his previous work (in 1889) on producing syntony between Leyden jars, as well as of its practical implications for wireless telegraphy [Lodge to Thompson, 14 April 1897, University College London (UCL)]. Those experiments had employed a species of closed circuit, which Lodge now adapted to the demands of wireless signalling, for which he filed a patent in May 1897 [Lodge 1897]. In his patent, Lodge described two principles. First, the damping of radiated waves was reduced by employing a closed-circuit transmitter, or a transmitter whose sparking surfaces were partially enclosed in a metallic box. Second, to receive a wave with a specific frequency, Lodge inserted a variable inductance into the receiver in series with the capacitance of the receiver's antenna.

At first, Marconi paid little attention to syntony. When he filed a provisional specification (June 1896) and a complete specification (March 1897) for his first wireless patent (No. 12039) he did not mention syntony at all [Marconi 1896]. Nor, at first, did syntonic issues much trouble him. Owing to the nature of Marconi's transmitter, the emitted wave was highly damped like Hertz's. As a result, the wave acted as if it were a mixture of waves having a broad range of frequencies, and this made his untuned receiving antenna respond extremely well. Before long, however, the market for wireless telegraphy forced Marconi to face up to syntonic issues. Initially, the demand for wireless telegraphy came from the military – the Army, the Navy, and the War Office – mainly because, unlike cable telegraphy, wireless signals did not travel over wires that the enemy could cut down. Marconi was well aware of this, and of the further fact that for military purposes secrecy was essential. And the only possible way to secure secrecy seemed to be syntony. In early 1898, Lodge's syntonic principle – the closed transmitter with variable inductance in the receiver – was made public by a paper Lodge read before the Physical Society in January 1898.[4] After this, talk about syntony became common. Both Adolf Slaby and Thompson pointed to Marconi's bad syntony as a serious flaw in his system for wireless telegraphy, and the London Times's long review of Marconi's system, as well as a leading article in the Electrician on the subject, agreed with them. All pointed out that heavy damping was a principal cause of bad syntony [Slaby 1898; Thompson 1898; London Times, 20 April 1898; (Leading Article), Electrician, 13 May 1898].

Marconi could not use Lodge's syntonic principle for two reasons. First, it was protected by patent. Second, and more importantly, it was not practical because the closed transmitter was not a strong radiator. In other words, though Lodge's closed transmitter was good for tuning, it was bad for practical communication because it could not transmit far. There was a fundamental and irreconcilable dilemma here: one needed an open circuit to produce powerful radiation, but one ended up as a result with a very dirty resonator; one needed a closed circuit to enforce good syntony, but this produced poor radiation. Marconi specifically wondered how to secure both characteristics at the same time. After many painstaking, and unsuccessful, trials in 1898–1900,[5] during which he experimented on adding an inductance or a capacitance to an antenna and even designed new antennas with exotic shapes (such as concentric cylinders), Marconi finally devised a practical syntonic system of his own. In this new system, a *closed* oscillatory circuit was connected to an *open* antenna by means of a special form of high-frequency oscillation transformer, the so-called jigger [Marconi 1900]. If one solves the series of complex differential equations that govern such a coupled circuit (as V. Bjerknes had done five years prior to Marconi's technological achievement [Bjerknes 1895]), one finds that the system, given a certain condition, generates nearly continuous waves.

Marconi's new syntonic scheme, which was also called the "four-seven" system after its patent number 7777 (1900), was not, however, free of problems. Worst of all, it was vulnerable to patent dispute since Lodge had also filed a patent on syntony. Ferdinand Braun in Germany had also applied in 1899 for a patent on "Improvements relating to the Transmission of Electric Telegraph Signals without Connecting Wires," which was publicized in January 1900 [Braun 1899]. Here, Braun, though without providing details, described a way of employing a transformer to connect a closed Leyden-jar discharge circuit with an oscillatory antenna. Like Braun, Marconi had employed a jigger for inductive coupling. Although Marconi mentioned that the antenna circuit "should preferably be suitably attuned for this purpose" and though he had specified a dimension for the transmitting jigger, Lodge's and Braun's patents rendered the basis of his 7777 claim vulnerable to challenge.

Marconi's scientific advisor, John Ambrose Fleming, became essential at this juncture. His scientific advisorship to the British Edison-Swan Company during 1882–1893 had mainly concerned patent matters, and Fleming had acted as an expert witness in other patent disputes. Just before Marconi filed the complete specification of his four-seven patent, Fleming, Marconi, John Fletcher Moulton (Marconi's patent attorney), and Flood-Page (Managing Director of the Marconi Company) discussed the best way to fortify the patent's claims. Envisioning litigation with Braun, Fleming asserted that the simplest strategy would be to argue that Braun's oscillatory transformer had itself been preceded by that of Nikola Tesla. But this was clearly not the best tactic, since it would also undermine the basis for Marconi's own claim [Fleming to Marconi, 20 Feb. 1901, Marconi Company Archives, (MCA)]. A superior strategy required emphasizing the novel aspects of Marconi's patent.

To that end Fleming singled out three points: first, Marconi's improvement consisted in the tuning of the four circuits in the transmitter and the receiver, requiring the condition $L_1C_1 = L_2C_2 = L_3C_3 = L_4C_4$, where the L_i, C_i are the respective inductances and capacitances; second, Marconi's four-fold tuning did, in fact, enormously increase the transmitting distance; third, to produce persistent vibrations in the transmitter the inductive coupling between oscillator and antenna had to be weak, not tight [Fleming 1901]. Fleming's recommendations were reflected in the Complete Specification, and, partly due to this "winning card" [Fleming to Marconi, 9 June 1911, MCA], the patent survived subsequent litigation.

MASKELYNE'S ATTACK AND THE CRUCIAL MARCONI–FLEMING EXPERIMENT

Marconi's four-seven system was adopted in designing a powerful transmitting station at Poldhu, Cornwall, for the experiment in transatlantic wireless telegraphy in 1901. Marconi's new syntonic system played a double role here. First, it increased the transmission distance. Only after Marconi had tuned all circuits of the Poldhu transmitter with his jiggers, did the transmission between Poldhu and Crookhaven in Ireland over 250 miles succeed [Hong 1996a]. Second, it helped persuade the members of the Board of the Marconi Company, who objected to risky experiments on the basis that the powerful waves required for transatlantic signalling would interfere with all the other ship–shore communications, on which the company's success resided. To persuade his board, Marconi had to demonstrate "isolating lines of communication" between the transmitting station at St. Catherine's and the receiving station at Poole, 30 miles away.

Three experiments were done using Marconi's four-seven arrangement. In the first, two transmitters, tuned to specific frequencies, sent different messages at the same time from St. Catherine's, and two separate receivers, respectively tuned to these transmitting frequencies, printed two distinct messages. Marconi then connected the two receiving aerials into one, and inductively linked his two receivers to this single antenna. Two messages – one in English and the other in French – were then sent from St. Catherine's, and the two receivers still vividly printed these two messages separately despite their connection to a single antenna. Finally, the transmission line between St. Catherine's and Poole was crossed by another line between Portsmouth and Portland, these two lines obliquely intersecting each other, but both signals were perfectly transmitted. These experiments were only performed before Marconi's board, but Fleming later wrote a letter to the *London Times* describing their success in commercial tuning. Fleming mentioned it again in an influential "Cantor Lecture" at the Society of Arts [Fleming 1900a, pp. 90–1; Fleming 1900b].

Marconi's system was not however altogether free from difficulties. Syntony proved to be troublesome at a New York yacht race in the summer of 1901, when Marconi's and De Forest's systems, both placed on racing boats to send messages to the press, generated interference with a then unknown, third system of wireless telegraphy [*Electrical World* **38**, 12 Oct. 1901, p.596]. Although this particular event did not have much influence, after Marconi's success in transatlantic wireless telegraphy in December 1901 syntony did become a significant public issue. Those who held financial and other interests in submarine cable telegraphy felt threatened, and they criticized Marconi's scheme. In the *Electrician* we read, e.g., the following:

If wireless communication can be established across the Atlantic, it would at all times be perfectly feasible for anyone either in England or in North America – or, indeed, anywhere over a far wider portion of the globe – to erect similar apparatus that would continuously and regularly "tap" every word of every message ... In fact, if Mr. Marconi establishes wireless telegraphy with America, his signals will be scattered all over Europe as well; ... everywhere within these boundaries they could be "tapped," and over this entire area they would interfere with other wireless telegraph apparatus working locally [(Leading Article) *Electrician*, 17 Jan. 1902].

This kind of criticism contained two different, but related, issues. One was a possibility of tapping, that is, loss of secrecy; the other was the possibility of interference with other working stations. The first problem also occurred in cable telegraphy and telephony. The second problem belonged mainly to wireless telegraphy.

Marconi addressed the fifth general meeting of the Marconi Company, held on 20 February 1902. After giving a brief description of experiments of the past two years on transatlantic telegraphy, and after refuting current scepticism about the accomplishment, Marconi confidently defended his syntonic system, stating that he could transmit across the ocean "without interfering with, or, under ordinary conditions, being interfered with, by any ship working its own wireless installations". He pointed out as well that an expert could tap the messages in cable telegraphy without cutting the cable. At the end of his speech, however, he included a rather strange and startling challenge:

I leave England on Saturday for Canada and expect to return to England about the end of March. If, on my return, either Sir W.H. Preece or Prof. Oliver Lodge, being of the opinion that, within the space of, say a week, after due notice given to me, he could succeed in intercepting and reading messages to be probably transmitted by me, at stated times, within that period between two of my stations, I should be happy to place any adjacent station of mine at his disposal for the purpose or if he should prefer to conduct his operations from a ship, he would, so far as I am concerned, be quite welcome to do so. [Marconi 1902, p.713]

The *Electrician* immediately responded that the terms of the challenge were not fair, because "it is not to be expected that either of these experts would be satisfied to use a neighbouring Marconi station, while it is unreasonable to expect them to build for themselves a somewhat costly station and with only a week for the task". It also pointed out that one week would be too short to tap Marconi's messages, supporting Marconi's statement. But this did not mean that the tapping was impossible. In principle, "syntony would be powerless to prevent any determined attempt to tap the signals which his stations scatter broadcast in all directions". In addition, it argued, the problem of interference was not at all addressed by proving difficulties in tapping. [(Leading Article) *Electrician*, 28 Feb. 1902].

As the *Electrician* predicted, neither Lodge nor Preece applied to take up the challenge. It was soon forgotten, buried beneath Marconi's other successes and scandals. During his trip to America in March 1902, he received *SSS* messages with a Morse printer across 1500 miles. This success was followed by harsh criticism from Silvanus Thompson, wherein Thompson revealed that neither the transmitting station at Poldhu nor the mercury coherer used at St. Johns was of Marconi's own design. Indeed, Thompson claimed that the transmitter was designed by Fleming, and the coherer had been invented by an Italian engineer named L. Solari. Thompson further (and not for the first time) argued that Oliver Lodge was the true inventor of wireless telegraphy [Thompson 1902; see also Hong 1994a].

Shortly after Thompson's attack, Marconi's attempt to file his own patent on Solari's coherer was revealed, and this raised questions concerning his ethics. Marconi escaped this difficulty by announcing the invention of a new magnetic detector based upon Ernest Rutherford's discovery of the magnetizing effect of electromagnetic waves in 1896 – a detector that freed Marconi altogether from the damaging effects of the coherer.[6] In the period from July to September of 1902, Marconi continued experimenting on long-distance wireless telegraphy between Poldhu and his sailing laboratory on board the "Carlo Alberto". While sailing to Italy in September, Marconi succeeded in receiving ordinary telegraphic messages (instead of a simple SSS) from Poldhu over a distance of 750 miles (1200 km), across both land and sea. This news, along with the first long-distance telegraphic messages, were widely publicized by Solari's report in the *Electrician* [Solari 1902]. Even the sceptical *Electrician* admitted that "Mr. Marconi is to be congratulated. [(Notes) *Electrician* 50 24 Oct. 1902]".

A strange comment, however, followed this congratulation. The *Electrician* stated that the signals had been tapped in England "with instruments not tuned by the Marconi Company for the purpose". Marconi's experiments, the article maintained, had in fact proved both the lack of secrecy in long-distance wireless telegraphy and the possibility of tapping at other stations. [Ibid.] The Managing Director of Marconi's Company quickly denied the *Electrician*'s claim, but the journal then published a detailed report on tapping by Nevil Maskelyne. Maskelyne published his Morse papers on which dots and dashes were printed, and claimed that the messages printed on his papers are exactly the same as what Solari had received aboard the "Carlo Alberto" sailing to Italy. He made other remarks which further startled Marconi and his company:

I had been almost constantly in touch with Marconi stations in various parts of the country. In every case I have found that our working produced mutual interference. Consequently, I inferred that those particular stations were not fitted with Mr. Marconi's syntonic apparatus. The reason for this I could not guess; but such was the fact. Of course, I had read of the marvellous efficacy of syntony. I had read of "triumphs" achieved almost every other day. I had read of experimentalists – even Sir Oliver Lodge, to whom we all owe so much – being challenged to intercept Marconi messages. Then the question arose, why does not Mr. Marconi use this syntonic apparatus? It seemed to be something too precious to be supplied to mere working stations. Still, he must use it *somewhere*, and the only conclusion at which I could arrive was that the syntonic arrangements must be employed exclusively at Mr. Marconi's latest and greatest station at Poldhu. Yet, when I went to Porthcurnow, 18 miles distant, I received Marconi messages with a 25 ft. collecting circuit raised on a scaffold-pole. No wonder I was interested. When, eventually, the mast was erected and a full-sized collecting circuit installed, the problem presented was, not how to intercept the Poldhu messages, but how to deal with their enormous excess of energy. That, of course, involved no difficulty, and by relaying my receiving instrument through landlines to the station in the valley below, I had all the Poldhu signals brought home to me at any hour of the night or day. It is for this reason that I claim to know something of the experiments conducted between Poldhu and the "Carlo Alberto". [Maskelyne 1902].

If Marconi's syntonic messages were so easy to intercept, then "what has become of that syntony of which we have heard so much"? Finally, Maskelyne returned a challenge to Marconi: "Can Mr. Marconi so tune his Poldhu station that, working every day and all day, it does not affect the station at Porthcurnow? ... he had only succeeded in proving that he cannot do so." [Maskelyne 1902].

Maskelyne, son of a wealthy and well-known family, was a self-educated electrician, who had been interested in wireless telegraphy since the late 1890s. In 1899, he demonstrated that one could explode gunpowder by wireless control, which generated widespread public interest. In 1900, he demonstrated a system at the annual meeting of the British Association, Bradford, to communicate between a balloon and the base, separated about 10 miles [London Times, 22 Sep. 1901]. With H.M. Hozier, secretary of Lloyd's, he had attempted to develop a wireless telegraphic system for the insurance company, but, without employing a grounded antenna, i.e., without violating Marconi's patent, they could not obtain practical results. After Lloyd's contracted for that purpose with Marconi, Maskelyne began to criticize Marconi's monopoly of wireless telegraphy and thus became a leading figure in the anti-Marconi faction. In 1907, he eventually became involved as an active member in the formation of the British De Forest Company. Maskelyne had several patents on wireless telegraphy, but after his debacle with Lloyd's, he seemed to be more involved in attacking Marconi than in developing his own practical system.[7]

Following Maskelyne's report, critical opinions concerning Marconi's Poldhu experiments poured forth. It was argued that "it is far more important that we should possess an effective ship-to-ship and ship-to-shore telegraphic system than that the Marconi Company should be able to establish telegraphic communication across the Atlantic. [(Leading Article) Electrician 7 Nov. 1902]". The Marconi Company castigated Maskelyne's printed messages as forgeries, and asserted that tapping by an expert was not in any case the issue since it was already possible in both cable telegraphy and telephony. Instead, it was argued, the most important thing for powerful transmission was to avoid interference with ordinary ship–shore communication, and the Marconi Company was confident that it could do so, and that it in fact had done so at Poldhu [Hall 1902]. In short, the company began now to emphasize interference rather than tapping. But the criticism did not stop. Critics requested a definite "proof of the justice of their repeated claim that the Poldhu station can be worked ... without interfering with other stations in its neighbourhood." [(Notes) Electrician 21 Nov. 1902].

After his return to England from America in February 1903, Marconi decided to produce a crucial "show" to prove the workability of his syntonic system as well as the safety of the Poldhu power station. The program was prepared by Fleming. A small transmitting station that simulated an ordinary ship-station was built 100 meters distance from the gigantic Poldhu station, the power of the former being 1/100th that of the latter. The frequency of this small station was tuned to a different frequency from that of Poldhu. Two of Marconi's duplex receivers, one tuned to the small station, the other to the gigantic Poldhu station, were installed at a station at the Lizard, a small town 6 miles from Poldhu. The experiment consisted in transmitting two different messages from these two stations at the same time and receiving them separately at the Lizard station.

Fleming wrote two kinds of messages, for a total of sixteen in all. Eight were typical ship-to-shore messages for the small station; the other eight consisted of cipher messages in ABC as well as several simple English messages. These were to be transmitted by the big Poldhu station. Keeping copies for later comparison with the transmitted signals, Fleming put the messages into sealed envelopes for secrecy and without showing them to Marconi. He inspected the Poldhu station to determine that it was working to its full capacity. To be certain that it was, Marconi arranged for another of his stations, this one located 200 miles away at Poole, simultaneously to receive the messages sent from the Poldhu station. Since Fleming was to inspect the messages received at the Lizard in the company of Marconi, he arranged a disinterested assistant, "unconnected with the Marconi organization," to remain at Poldhu where his duty was to open the envelopes at a certain time and to deliver the first two messages to the operators of the two stations. The operators were to send these messages for ten minutes, and then, after a five minute interval, the assistant was to give them the envelopes containing the second two messages, and so on. The experiment was conducted on 18 March 1903 from 2 p.m. to 4 p.m. Marconi and Fleming watched the messages print at the Lizard station. Afterwards Fleming asked Marconi to read the messages that they thought to have received from the small station. Marconi read off and wrote down the eight messages "without a single mistake".

In the Cantor Lecture on 23 March and in a long letter to the *London Times* on 4 April, Fleming proudly claimed these results as a complete demonstration of the lack of interference between Marconi's syntonic systems [Fleming 1903a, p. 772; Fleming 1903b]. At the sixth general meeting of the company, held on 31 March 1903, Marconi had declared that Fleming's experiments confirmed his belief that no interference would occur between the Poldhu station and ordinary ship–shore communications, provided that they were all equipped with Marconi Company equipment [Marconi 1903]. Marconi's and Fleming's triumph did not go long unchallenged.

MASKELYNE INTERFERES WITH FLEMING'S ROYAL INSTITUTION LECTURE
(4 JUNE 1903)

The Marconi–Fleming experiment had not been a public demonstration, nor had an unimpeachable witness inspected the received messages, which had after all been read by Marconi and compared with copies kept by Fleming. The authenticity of the demonstration, and thus of the syntony and safety of Marconi's system, relied entirely upon Fleming's own testimony [for a discussion of witnessing and credibility in early modern science, see Shapin and Schaffer 1985]. His authority resulted from the high credit he had accrued over the previous two decades within the British electrical engineering and physics community. Fleming was known as an exact, honest, and hard-working scientist–engineer, one who had laid the foundation of scientific electrical engineering [Hong 1995a; 1995b]. Could Fleming's prior credit be regarded as valuable in the new world of wireless telegraphy?

Two differences between the old and the new worlds made doing so difficult. First, different kinds of "ether engineers" – physicists like Lodge, power engineers like Fleming, telegraphers like A. Muirhead and W.H. Preece, and self-educated ones like Marconi and Maskelyne – were now competing with each other for authority in the novel world of "ether engineering". Second, as a result of Marconi's own virtual monopoly of the market, the challenge to him was extraordinarily intense and came from several directions. As a result, from 1899 to 1903, Fleming had had to work very hard indeed to gain credibility in this new field, and he had done so almost entirely by translating Marconi's practical wireless telegraphy into the terms of scientific engineering. His two Cantor lectures in 1900 and 1903 [Fleming 1900a; 1903a] were read worldwide, and they were cited as providing the first scientific rationale for wireless technologies. Fleming had now exploited his authority in publicizing, as a qualified witness, Marconi's secret demonstrations. To Marconi's opponents, therefore, challenging Fleming's credibility became essential.

Just after Fleming announced Marconi's success in practical syntony, Maskelyne requested that "we [...] have a right to demand the absolute justification of his [Fleming's] claims" [Daily Telegraph, 25 March 1903]. Several days later, at a Company meeting, Marconi stated that Fleming would repeat the experiment under the auspices of Lord Kelvin and Lord Rayleigh [Marconi 1903]. This repetition was not however performed. Instead, Fleming prepared laboratory-scale demonstrations to be given at the Royal Institution in a pair of public lectures entitled "Electric Resonance and Wireless Telegraphy". The first lecture, which was given on 28 May, mainly concerned theoretical aspects of Hertzian resonance and their practical implications for wireless telegraphy [Electrician, 29 May 1903, p. 235]. In the second lecture, held on 4 June, Fleming first discussed conditions for tuning, as well as for the generation and reception of syntonic oscillations. According to the very brief summary that appeared two weeks later in the Electrician [12 June 1903, p. 315], Fleming "referred to the possibility of tuning transmitters and receivers ..., pointing as an instance, to the experiments recently carried out by Mr. Marconi on the south coast of England". Then, the Electrician continued, Fleming gave demonstrations of reception of two wireless messages, one being sent out from Fleming's laboratory at University College London, and the other from Marconi himself at the distant station in Poldhu. These latter were not, it was noted, received directly at the Royal Institution, but had rather been received first at Chelmsford (where Marconi had his factory), near London, from which they were retransmitted; the reason for this contrivance lay in the claim that the antenna erected temporarily at the Royal Institution was simply too short (60 feet) to receive the Poldhu message directly. These demonstrations were not designed to show lack of interference between different systems transmitting simultaneously at different frequencies, but rather to demonstrate the power of wireless signalling with such a tuned system. But Maskelyne had other goals in mind.

According to his recollection, Maskelyne had at first planned to attend the lectures, but "at once grasped the fact that the opportunity was too good to be missed". He decided to test Marconi's claim of non-interference by trying himself to interfere in Fleming's lecture on syntony, which, he felt, was "something more than a right; it was a duty". Marconi had always claimed that even the gigantic Poldhu station would not interfere with his low-power ship-to-shore communication, and, if this were really the case, Maskelyne reasoned, ordinary transmitting stations should certainly not trouble communications between Marconi stations. He thus installed his 10″ induction coil at a theatre (the "Egyptian Theatre", owned by his father) near the Royal Institution, and adjusted his transmitter so that the radiation was "out of tune" with Marconi's. To do this, he exploited short waves, as Marconi was known to use long waves. The final problem was how to check whether his trial was successful or not. For this he sent messages "calculated to anger and draw somebody at the receiving end" [see the recollection in Maskelyne 1903a, p. 358]. The lecture began at 5 p.m., and, after some time (around 5:45 p.m.), Maskelyne started transmitting.

Thanks to Arthur Blok, one of Fleming's assistants, who assisted him at the lecture, we have a detailed description of what happened in the lecture theatre of the Royal Institution:

One of the Marconi Company's staff was waiting at the Morse printer and while I [Arthur Blok] busied myself with demonstrating the various experiments I heard an orderly ticking in the arc lamp of the noble brass projection lantern which used to dominate this theatre like a brazen lighthouse. It was clear that signals were being picked up by the arc and we assumed that the men at Chelmsford were doing some last-minute tuning-up.

But when I plainly heard the astounding word "rats"[8] spelt out in Morse the matter took on a new aspect. And when this irrelevant word was repeated, suspicion gave place to fear. The man at the printer switched on his instrument and the hands of the clock moved inexorably towards the minute when the Chelmsford message was to come through. There was but a short time to go and the "rats" on the coiling paper tape unbelievably gave place to a fantastic doggerel, which, as far as my memory serves me, ran something like this

> There was a young fellow of Italy
> Who diddled the public quite prettily

A few more lines en suite completed the verse, and quotations from Shakesphere also came through.

Evidently something had gone wrong. Was it practical joke or were they drunk at Chelmsford? Or was it even scientific sabotage? Fleming's deafness kept him in merciful oblivion and he calmly lectured on and on. And the hands of the clock, with equal detachment, also moved on, while I, with a furiously divided attention, glanced around the audience to see if anybody else had noticed these astonishing messages. All seemed well – a testimony to the spell of Fleming's lecture – until my harassed eye encountered a face of supernatural innocence and then the mystery was solved. The face was that of a man whom I knew to be associated with the late Mr. Nevil Maskelyne in some of his scientific work.

The story ended happily, at least as far as the lecture was concerned. By a margin of seconds before the appointed Chelmsford moment, the vagrant signals ceased and with such *sang froid* as I could muster I tore off the tape with this preposterous dots and dashes, rolled it up, and with a pretence of throwing it away, I put it in my pocket. The message from Chelmsford followed and Fleming's luck had held amidst much unsuspecting applause. [Blok 1954]

Three points in this recollection should be noted. First, we see that neither Fleming nor Marconi had the slightest expectation of any attempt to ruin their public show. Fleming designed his public demonstrations with the utmost care, and he was confident in the efficacy of Marconi's system. Second, Marconi's antennas and receivers at the Royal Institution were supposed not to pick up any wireless messages other than Marconi's own, because they were designed on the basis of Marconi's four-seven system. Finally, Fleming received the Poldhu messages (relayed by Chelmsford), not because of a lack of interference from Maskelyne's messages, but because the latter's messages had stopped before the ones from Chelmsford arrived. Had Maskelyne's messages continued for a few more minutes, they would have certainly generated jumbled results by mixing with Marconi's.

Fleming was told about the attempt after the lecture, and was infuriated. The next day, on 5 June 1903, he reported to Marconi that "there was however a dastardly attempt to jamb us ... I was told that Maskelyne's assistant was at the lecture and sat near the receiver. Nothing of the kind happened at previous rehearsals but at 5:45 p.m. Mr Woodward [an engineer of the Marconi Company] got a message down which came 'through the ether' from somewhere and there were mysterious effects heard in the arc lamp which seemed to indicate 'electric jerks' as Lodge would say being put with the earth plate or the aerial. Maskelyne's assistant had been seen loafing about some days before. Anyway the attempt was not successful in spoiling our 'show.' We got of course excellent signals from my laboratory." [Fleming to Marconi, 5 June 1903, MCA]. We do not know whether Marconi replied to Fleming, since no such letter has survived, but we do have Fleming's next letter to Marconi in which he considered the attempt to have been done by flowing a strong earth current nearby, and continued: "Professor Dewar [the director of the Royal Institution] thinks I ought to expose it. As it was a purely scientific experiment for the benefit of the R.I. it was a ruffianly act to attempt to upset it, and quite outside the 'rules of the game'. If the enemy will try that on at the R.I. they will stick at nothing and it might be well to let them know" [Fleming to Marconi, 6 June 1903, MCA].

Why was it outside the rules of the game? Fleming apparently believed that someone had created a strong earth current to destroy the grounding of Marconi's antenna, thereby interfering with syntonic communication in an unacceptable, utterly unusual, fashion. In other words, he had complete confidence in the efficacy of Marconi's four-seven system. Yet, contrary to

Fleming's conjecture, Maskelyne had not in fact used an earth current; he had used electromagnetic waves.[9] Fleming was wrong, but why had he been so mistaken? His mistake lay in assuming that if Maskelyne *had* actually sent out proper air waves, then he would certainly have used a syntonic transmitting system similar to what Fleming and Marconi were employing. In other words, anyone seeking to do so would have sought to interfere with Marconi by himself using a narrowly defined, but different, frequency from the one being used to transmit from Chelmsford and from University College. Marconi's four-seven devices were, Fleming was utterly convinced, protected from interference with each other, and so if Maskelyne had succeeded it could not have been by using air waves. It apparently never occurred to Fleming that Maskelyne would use a simple induction coil to generate short waves. Maskelyne, perhaps entirely unintentionally, had exploited "dirty" waves, i.e., waves having a broad frequency range, and against such things Marconi's syntonic system was utterly vulnerable. The old, simple technology of the spark-gap transmitter, which Fleming and Marconi, along with many others, had long ago thrown into the trashbin, now trumped them. Fleming could not imagine Maskelyne to have used such an inherently faulty and outdated technology, and so he concluded that Maskelyne must have deliberately tricked him in the simplest way possible, which was by destroying the efficacy of Fleming's grounded antennas by sending a strong current directly to earth from a nearby hiding-place. Maskelyne, for his own part, probably did not intentionally employ a device that he knew ahead of time would necessarily be detected by any device whatsoever, no matter how it was designed. Indeed, even after the interference he seems not fully to have understood the reason for his success.[10]

On 8 June, Fleming wrote a letter to the *London Times* (published on 11 June), in which he stated that a "deliberate attempt" to "wreck the exhibition" of London–Poldhu communication had been undertaken by "a skilled telegraphist and someone acquainted with the working of wireless telegraphy". Fleming then emphasized the scientific nature and purposes of his own demonstration by invoking the sacred name of Faraday and Faraday's theatre at the Royal Institution, bemoaning the failure that "the theatre which has been the site of the most brilliant lecture demonstrations for a century past" was not protected from the "the attacks of a *scientific hooliganism*". He then begged of the reader any clue to the names of those responsible for these "monkeyish pranks" [Fleming 1903c]. Fleming's *Times* letter led the *Morning Leader* quickly to interview Cuthbert Hall, the managing director of the Marconi Company. Hall stressed that, "in spite of all", the Poldhu messages had nevertheless gotten through ["Wireless Rats" *Morning Leader*, 12 June 1903].

In a letter to the *London Times*, Nevil Maskelyne, who was anxiously waiting for Fleming's response, frankly and proudly admitted what he and his friend Dr. Horace Manders had done. However, Maskelyne denied Fleming's allegation that he had attempted to "wreck" the show. He defended himself on the basis that he could have easily done so, had he wanted to. Instead, he had in any case transmitted only for a few minutes. He justified his behaviour on two grounds: first, what Fleming attempted in the lecture was to show "the reliability and efficacy of Marconi syntony," which was controversial, rather than scientific principles of Hertzian syntony, which were not; second, Maskelyne had adopted "the only possible means of ascertaining *fact* which ought to be in the possession of public". What was this fact? It was: "a simple untuned radiator upsets the 'tuned' Marconi receivers" [Maskelyne 1903b].

After Maskelyne's confession, the *Daily Express* interviewed Fleming. Evidently taking some pleasure in having identified Maskelyne, Fleming remarked that he "had anticipated some attempts of the kind" and thus "took some suitable precautions". Ignoring or just overlooking Maskelyne's reference to his "simple untuned radiator", Fleming repeated his belief that "there was clear proof that somebody quite close was putting high-tension currents into the earth", which made the attempt "not a fair interference" or "getting in at the back door". Yet, in spite of all this, "there was no actual success in interfering with my lecture". Finally, he again referred to the "scientific lecture" in the "sacred" Royal Institution that "should be kept free from things of this sort" ["Ghost's Tapping" *Daily Express*, 13 June 1903].

It is interesting to notice that Fleming invoked the names of science and of the Royal Institution in blaming Maskelyne [Fleming 1903d], whereas Maskelyne defended himself in the name of the public. This claim of course depended to a considerable extent on the implicit argument that Fleming had not been dealing specifically with Marconi's syntony, or that he had been speaking about physical principles that transcended the particularities of the commercial world. This was hardly compelling to someone like Maskelyne, and probably to many others as well. The receiving antenna and the other apparatus in the Royal Institution had been installed with Marconi's help, and it was also he who sent messages from Poldhu for Fleming's lecture. Further, Fleming had definitely stated in the lecture, as one observer later recalled, that "his instruments were so arranged that no other person could prevent messages from Poldhu from reaching him" [Black 1903].

Maskelyne, in reply, wrote that "the Professor called up the name of Faraday in condemning us for what we did. Supposing Faraday had been alive, whom would he have accused of disgracing the Royal Institution – those who were endeavouring to ascertain the truth or those who were using it for trade purposes? Professor Fleming gave two lectures that afternoon. The first was by Professor Fleming, the scientist, and was everything that a scientific lecture ought to be; the second was by Professor Fleming, the expert advisor to the Marconi Company" ["Scientific Hooliganism" St. James's Gazette, 13 June 1903; Maskelyne 1903c]. To many Fleming's name was by then so intimately connected to that of Marconi that it was difficult to associate him with pure Hertzian physics, and in any case it was commercial telegraphy, not the rarefied realm of "ether waves," that bore on public interests. In a letter to the *London Times*, Charles Bright, an old and respectable telegrapher, requested a "full, open, and disinterested inquiry as to the merits of the system from every standpoint, accompanied by a complete series of trials by impartial authorities" [Bright 1903]. Following this, the *Electrical Review* took Maskelyne's side, criticizing Fleming in an editorial, entitled "Who was the Hooligan"?:

If it turns out that Mr. Maskelyne made use of extraordinary means to upset the lecturer, Prof. Fleming has some grounds for his protest; but if, as we think, the means employed were fair, and such as might occur in practice, then the protest must be made by the public against the professor. In his dual capacity of *savant* at a learned institution, and expert to a commercial undertaking, Prof. Fleming is discovering that while the public have nothing but courtesy and respect to offer to the one, they have searching criticism and still more searching experiment to oppose, if need be, to the statements of the other. When the philosopher stoops to commerce he must accept the conditions of commerce. Having descended from the high pedestal, upon which science placed him, into the arena of trade and competition, he attempts, when attacked, to climb back to the sacred temple, whence, like the proverbial armadillo, he can in safety evade, observe and deride his pursuers. But, as a matter of fact, it is ridiculous for him to appeal in his dilemma to the reverence which we all feel for the traditions of the Royal Institution. Faraday, to begin with, would never have placed himself in so anomalous a position; and, moreover, Faraday would have displayed more interest in what Mr. Nevil Maskelyne was trying to uncover, than in what, if we are to believe Mr. Maskelyne's statement, Prof. Fleming was seeking, for commercial reasons, to withhold from public knowledge. [(Editorial), *Electrical Review*, 19 June 1903].

The controversy terminated in an unexpected way. On 10 July, in a column of the *Electrician*, Maskelyne cast doubt on whether the Poldhu station itself, from which Marconi had sent his message to Chelmsford on 4 June, had actually sent the message during Fleming's lecture, i.e. from 5 to 6 p.m. Maskelyne allegedly argued that only one message, "BRBR. Best Regards to Prof. Dewar sent through ether from Poldhu. - Marconi. PDPD", was sent from 11:50 a.m. only for several hours [Maskelyne 1903d]. In the next issue of *Electrician* (17 July 1903), Maskelyne admitted that he was in fact wrong about this, since another station 50 miles apart from Poldhu captured a message repeatedly sent out from the Poldhu station – a message concerning the reestablishment of transatlantic communication – after 5 p.m. i.e., during Fleming's lecture. Maskelyne however appended, without giving further details, that "its wording was not exactly that of the message received at the Royal Institution" [Maskelyne 1903e]. The Marconi company replied to Maskelyne's previous (July 10) attack by publishing a detailed time table of the working of both the Poldhu and the Chelmsford stations on 4 June, as well as a copy of the message sent from them, which was endorsed by the signatures of the four technicians (two technicians each) at these two stations, as well as by those of Fleming and his assistant at the Royal Institution. The Company's rebuttal certified that the message, which read "My best thanks to President Royal Society and yourself for kind message; Communication from Canada was reestablished May 23rd", had indeed been sent from Poldhu to Chelmsford at 5:15 p.m., and relayed from there to the Royal Institution at 5:25 p.m. [Allen 1903].

Unfortunately for Fleming and Marconi, the specific message which the Company certified had been sent out from Poldhu did not tally well with the one that had been received by Fleming at the Royal Institution. According to Maskelyne's subsequent exposure, the message that Fleming received and read at the Royal Institution specifically ran "P.D. to R.I. C To Prof. Dewar. To

President Royal Society and yourself thanks for kind message; Communication from Canada was re-established May 23rd – Marconi" [Maskelyne 1903f]. The messages clearly had the same meanings, but, as Maskelyne had said, the wording was nevertheless quite different. Fleming had undoubtedly received some message from Chelmsford since it was taken down on the spot, but it was apparently not the same one that the Marconi company had sent out from Poldhu. Given the evidence that the message sent out from Poldhu was indeed captured at another station near it, there remain two possibilities for what happened. First, perhaps technicians at Chelmsford, while dispatching the message, had somehow mis-transcribed the one from Poldhu. However, it is very unlikely, though not impossible, that skilled technicians would mis-transcribe such a simple message. The second possibility is that the message sent from Poldhu had not in fact been received, or had been very imperfectly received, in Chelmsford. In this case, the message that Fleming received at the Royal Institution would simply have originated at Chelmsford – by someone who had somehow known the content of Marconi's congratulatory message. Maskelyne had few doubts about what had probably happened. As he put it, "Chelmsford could easily have sent such a message without having received a word from Poldhu". After this, Fleming and Marconi stopped responding. The controversy thus ended, but the entire authenticity of Fleming's scientific show was cast into doubt. This was a serious blow to Fleming as well as to Marconi's Company.

The Maskelyne affair did considerable damage to Fleming's credit. Afterwards he could no longer claim to be a "qualified witness" where Marconi was concerned. He could not, and did not, write any further letters to the *London Times* concerning the matter. There was more. *Punch* sneered at Marconi, writing of a daily newspaper on the transatlantic boat produced by means of "Marconigrams" and resulting in topsy-turvy messages, because of interference with a nearby cruiser ["The Daily Wireless", *Punch* **124**, 1 July 1903, 453].

The Maskelyne affair enforced Marconi to moderate his love for show and public demonstration. Since 1896 Marconi had used shows to attract attention and to exhibit the practical nature of his wireless telegraphy, shows performed for and by famous people including Lord Kelvin, the poet Tennyson, the Prince and other members of the Royal Family, the Italian King, the Russian Emperor, the American President, and many more. Published reports by the Marconi Company of these performances glowingly described success after success. Yet the actual performances were at times rather different from the Company's portrayal of them. In September 1902, for example, Marconi had been keen to receive on board the "Carlo Alberto" congratulatory messages sent from Poldhu for the Italian King. The Solari's published report of the event stated that this had gone quite smoothly. In fact reception had been extremely difficult, so bad in fact that the temperamental Marconi had at one point smashed all the receivers on board. During his second transatlantic experiments in the winter of 1902–03, signal transmission and reception of signals had been altogether too difficult and unstable, though the published papers described the finals results as if they had been entirely straightforward, indeed simple and quite easy, because of Marconi's superior system [D. Marconi 1970, pp. 116-123].

THE CYCLE OF FLEMING'S CREDIBILITY

The Maskelyne affair illuminates several interesting aspects in the early history of wireless telegraphy. First of all, it betrays a major, indeed critical, limitation of Marconi's syntonic system. Marconi's, or any, syntonic system was extremely vulnerable to the old, simple transmitters that generated "dirty" waves. In 1901, Marconi had declared that "the days of the non-tuned system are numbered" [Marconi 1901, p. 515], and Fleming had at the time confidently remarked that "it will now be possible to provide the Admiralty and the Post office with instruments having an Admiralty or a Post Office frequency, and to register frequency just as a telegraphic address is registered, so that no one else could use that particular frequency" [mentioned in (Leading Article), Electrician, 24 May 1901]. The Maskelyne affair pointed out that interference with other communication could only be avoided by a sort of instrumental monopoly under which unsyntonized transmitters would be strictly prohibited from existing at all. The affair reflected, and accelerated, the shift of wireless technology from a period during which public shows and sensations (such as the first transatlantic reception of SSS signals) had been essential to economic success, to one in which the regulation of frequencies and the guarantee of instrumental uniformity became a serious issue.

Fleming had been appointed scientific advisor to the Marconi Company in 1899 for a year with an annual salary of £300.[11] In December 1900, when he was fully engaged with Marconi's first efforts at transatlantic telegraphy, Fleming successfully increased his wages to £500 a year for three years [Hong 1996a]. From 1899 to 1903, Fleming's role was limited mainly to helping Marconi create public sensations and to reporting Marconi's private shows to the British public. The Maskelyne affair strikingly devalued his credibility, and this immediately prevented him from working any longer as a trusted witness. Moreover, his three-year contract with the Marconi Company, which terminated in December 1903, was not renewed, for Fleming had apparently become useless to Marconi. Seeking to revive his relation with Marconi, Fleming resorted to his Pender Laboratory in University College London, where, in 1904, he invented a wavelength measuring instrument, the Cymometer, and a new, high-frequency alternating current rectifier, the valve. With these devices in hand he successfully reestablished a Marconi connection May 1905, having, as it were, created new technical capital and, in direct consequence, a new source of credibility [Hong 1994b; 1996b].

Institute for the History and Philosophy of Science and Technology, University of Toronto, Canada

REFERENCES

Aitken, G.J. Hugh. 1976. *Syntony and Spark: The Origins of Radio*. New York.

Allen, Henry W. 1903. "Wireless Telegraphy at the Royal Institution: To the Editor of *The Electrician*". *Electrician* 51 *(17 July): 549*.

Austin, L.F. 1903. "Our Note Book". *The Illustrated London News* 122 (20 June).

Bjerknes, V. 1895. "Ueber Electrische Resonanz". *Wiedemann's Annalen* 55 (1895), 121–69.

Black, George. 1903. "Wireless Telegraphy at the Royal Institution: To the Editor of The *Electrician*". *Electrician* 51 (10 July): 503.

Blok, Arthur. 1954. "Some Personal Recollections of Sir Ambrose Fleming," (The third Fleming Memorial Lecture, Royal Institution, 29 September 1948), published as Appendix II of J.T. MacGregor-Morris, *The Inventor of the Valve: A Biography of Sir Ambrose Fleming*, pp. 124–34. London: The Television Society.

Braun, F. 1899. "Improvements relating to the Transmission of Electric Telegraph Signals without Connecting Wires". British Patent No. 1862. Date of application, 26 January 1899; Complete specification, 23 October 1899; Accepted, 6 January 1900.

Bright, Charles. 1903. "To the Editor of *The Times*". *London Times* (16 June).

Buchwald, Jed Z. 1994. *The Creation of Scientific Effects: Heinrich Hertz and Electric Waves*. Chicago.

Buchwald, Jed Z. forthcoming. *The Creation of Scientific Effects II*.

Dam, H.J.W. 1897. "Telegraphing Without Wires: A Possibility of Electrical Science," (interview with Marconi). *McClure's Magazine* 8 (March): 383–92.

[Editorial]. 1903. "Who was the Hooligan". *Electrical Review* 52 (19 June): 1031.

Fleming, J.A. 1900a. "Electrical Oscillations and Electrical Waves," (Cantor Lecture). *Journal of the Society of Arts* 49: 69–131.

Fleming, J.A. 1900b. "Recent Advances in Wireless Telegraphy: To the Editor of *The Times* ". *London Times* (4 October).

Fleming, J.A. 1901. "A Few Notes on No. 7777 of 1900," (written in February, 1901), Marconi Company Archive, Chelmsford.

Fleming, J.A. 1903a. "Hertzian Wave Telegraphy," (Cantor Lecture). *Journal of the Society of Arts* 51: 709–84.

Fleming, J.A. 1903b. "Power Stations and Ship-To-Shore Wireless Telegraphy: Letter to the Editor of *The Times*". *London Times* (14 April).

Fleming, J.A. 1903c. "Wireless Telegraphy at the Royal Institution: To the Editor of *The Times*". *London Times* (11 June).

Fleming, J.A. 1903d. "Wireless Telegraphy at the Royal Institution: To the Editor of *The Times*". *London Times* (16 June).

Hall, H.C. 1902. "Wireless Telegraphy: To the Editor of The *Electrician*". *Electrician* 50 (21 November): 198–9.

Hong, Sungook. 1994a. "Marconi and the Maxwellians: The Origins of Wireless Telegraphy Revisited". *Technology and Culture* 35: 717–49.

Hong, Sungook. 1994b. "From Effect to Artifact: The case of the Cymometer", *Journal of the Korean History of Science Society* 16: 233–49.

Hong, Sungook. 1995a. "Efficiency and Authority in the 'Open versus Closed' Transformer Controversy". *Annals of Science*, 52: 49–76.

Hong, Sungook. 1995b. "Forging Scientific Electrical Engineering: John Ambrose Fleming and the Ferranti Effect". *Isis*, 86: 30–51.

Hong, Sungook. 1996a. "Styles and Credit in Early Radio Engineering: Fleming and Marconi on the First Transatlantic Wireless Telegraphy". *Annals of Science* 53 (forthcoming).

Hong, Sungook. 1996b. "From Effect to Artifact (II): The Case of the Thermionic Valve". *Physis* (forthcoming).

Hong, forthcoming. "From Hertz to Marconi's Telegraphy: The Laboratory and the Field in Early Wireless Experiments, 1888–1896". (A draft version is available from the author).

Jolly, W.P. 1972. *Marconi*. London.

[Leading Article], "Wireless Telegraphy". *Electrician* 41 (13 May 1898): 82–3.

[Leading Article], "The Tuning of the Wireless Telegraphy". *Electrician* 47 (24 May 1901): 176–7.

[Leading Article], "The Limitation of Wireless Telegraphy". *Electrician* 48 (17 January 1902): 500–1.

[Leading Article], "A Reply to Mr. Marconi". *Electrician* 48 (28 February 1902): 732–3.

[Leading Article], "Practical Wireless Telegraphy". *Electrician* 50 (7 November 1902): 102–3.

Lodge, O. 1894. "Work of Hertz," (Friday Evening Lecture). *Proceedings of the Royal Institution* 14: 321–49.

Lodge, O. 1897. "Improvements in Syntonized Telegraphy without Wires". British Patent No. 11575. Date of application, 10 May 1897; Complete specification, 5 February 1898; Accepted, 10 August 1898.

[Lodge, O]. 1898. "Dr. Lodge on Wireless Telegraphy". *Electrical Review* 42 (28 January): 103–4.

[London Times]. 1898. "Wireless Telegraphy". *London Times* (20 April).

Marconi, G. 1896. "Improvements in Transmitting Electrical Impulses and Signals, and in Apparatus Therefor". British Patent No. 12039. Date of application, 2 June 1896; Complete specification, 2 March 1897; Accepted, 2 July 1897.

Marconi, G. 1900. "Improvements in Apparatus for Wireless Telegraphy". British Patent No. 7777. Date of application, 26 April 1900; Complete specification, 25 February 1901; Accepted, 13 April 1901.

Marconi, G. 1901. "Syntonic Wireless Telegraphy," (Cantor Lecture). *Journal of the Society of Arts* 49: 506–15.

Marconi, G. 1902. "Address," (Fifth general meeting of the Marconi's Wireless Telegraph Company, held on 20 February 1902). *Electrician* 48 (21 February): 712–13.

Marconi, G. 1903. "Address," (Sixth general meeting of the Marconi's Wireless Telegraph Company, held on 31 March 1903). *Electrician* 50 (3 April): 1001–2.

Marconi, Degna. 1970. *My Father Marconi*. Ottawa.

Maskelyne, N. 1902. "A Supplement to Lieut. Solari's Report on 'The Radio-Telegraphic Expedition of H.I.M.S. Carlo Alberto'". *Electrician* 50 (7 November): 105–9.

Maskelyne, N. 1903a. "Electrical Syntony and Wireless Telegraphy". *Electrician* 51 (19 June): 357–60.

Maskelyne, N. 1903b. "Wireless Telegraphy: To the Editor of *The Times*", *London Times* (13 June).

Maskelyne, N. 1903c. "Wireless Telegraphy at the Royal Institution", *London Times* (18 June).

Maskelyne, N. 1903d. "Wireless Telegraphy at the Royal Institution", *Electrician* 51 (10 July): 503.

Maskelyne, N. 1903e. "To the Editor of the Electrician". *Electrician* 51 (17 July): 549.

Maskelyne, N. 1903f. "Wireless Telegraphy at the Royal Institution", *Electrician* 51 (24 July): 592.

[Notes]. *Electrician* 50 (24 October 1902): 54.

Pocock, Rowland F. 1988. *The Early British Radio Industry*. Manchester.

Shapin, Steven and Simon Schaffer. *Leviathan and the Air-Pump: Hobbes, Boyle and the Experimental Life*. Princeton Univ. Press. 1985.

Slaby, Adolf. 1898. "The New Telegraphy: Recent Experiments in Telegraphy with Sparks". *Century Magazine* 55 (April): 867–74.

Solari, L. 1902. "The Radio-Telegraphic Expedition of H.I.M.S. Carlo Alberto". *Electrician* 50 (24 October): 22–6.

Susskind, Charles. 1968–1970. "The Early History of Electronics". *IEEE Spectrum* 5 (8): 90–98; 5 (12): 57–60; 6 (4): 69–74; 6 (8): 66–70; 7 (4): 78–83; 7 (9): 76–84.

Thompson, Silvanus P. 1898. "Telegraphy Across Space," (Cantor Lecture). *Journal of the Society of Arts* 40: 453–60.

Thompson, Silvanus P. 1902. "The Inventor of Wireless Telegraphy". *The Saturday Review* 93 (5 April): 425.

NOTES

I thank Bill Aspray, Jed Buchwald (especially), Hasok Chang, Peter Galison, Yves Gingras, Takehiko Hashimoto, Youngran Jo, Janis Langins, Trevor Levere, Rik Nebeker, and Roy Rodwell for their reading of various versions of this paper, as well as for helpful comments and insightful criticism. An earlier version of this paper was presented at the annual meeting of the Society of the History of Technology in Lowell, and at the colloquium of the Institute for the History and Philosophy of Science and Technology, University of Toronto, in 1994.

[1] The coherer, the earliest detector of wireless messages, was based on Edouard Branly's discovery in 1890 that fine copper filings, capsuled into a glass tube, were conducting only feebly

under normal conditions but their conductivity was greatly increased when a spark (thus Hertzian waves) was generated nearby. Its sensitivity was improved by Oliver Lodge in the early 1890s. For the story of the coherer, see Hong 1994a, p.725–6.

[2] Several pioneers in the period of 1888–95 either envisioned the possibility of using Hertzian devices for telegraphic communication (such as William Crookes) or performed some preliminary experiments for this purpose (such as Captain H. Jackson). The work of these scientists and engineers in this transition period between Hertzian physics and Marconi's telegraphy has been of interest to historians [Aitken 1976; Pocock, 1988; Susskind 1968–70], but the question of why Marconi succeeded where all others failed has not been satisfactorily answered. I have examined this question and early works of Crookes, Jackson, O. Lodge, Chunder Bose, A.P. Trotter, F.J. Trouton and E. Rutherford in the context of the discontinuous evolution of Hertzian devices into Marconi's telegraphy in my forthcoming article "From Hertz to Marconi's Telegraphy: The Laboratory and the Field in Early Wireless Experiments, 1888–1896.

[3] The phenomenon of multiple resonance was first discovered by two Swiss physicists, E. Sarasin and Lucien de la Rive. On the basis of their discovery, they claimed that the wave generated by the Hertzian device was a composite of heterogeneous waves (as white light is a composite of various heterogeneous waves) dispersed throughout a very broad frequency range. Hertz attributed this anomaly to damping of a wave with a definite frequency. Multiple resonance is discussed by Aitken [Aitken 1976, pp. 70–3]. Buchwald's forthcoming monograph on the replication of Hertzian effects will also discuss this phenomenon [Buchwald forthcoming].

[4] Lodge's paper was not published in the *Proceedings of the Physical Society*, but it was fully described in Lodge 1898.

[5] These early experiments played a significant role in the design of his successful four-seven system, since Marconi was able to discover an empirical rule for his jigger windings that secured the best syntonic conditions [see Marconi 1900]. Such an empirical rule was very important, because at the time it was impossible to satisfy the syntonic condition, $L_1 C_1 = L_2 C_2$, by measuring or calculating the capacitance and inductance of the various circuits. The primary problem was measuring inductance. Marconi once recalled [Marconi 1901, p. 511] that "I have found it impracticable by any of the methods with which I am acquainted directly to measure the inductance of, say, two or three small turns of wire. As for calculating the inductance of the secondary of small transformers, the mutual effect of the vicinity of the other circuit and the effects due to mutual induction greatly complicated the problems". Achieving syntony at that time was not a matter of calculation; it was a *craft*. Marconi was the one who first mastered it with the jigger.

[6] This announcement was quite unusual, since the patent for the magnetic detector was not fully granted. Marconi had filed only a Provisional Specification in May 1902. Prior to this time, Marconi had never publicized his invention before his complete specification had been accepted.

[7] I was not able to locate a reliable biographical source on Nevil Maskelyne. The information in this paragraph is collected from various engineering journals and newspapers of the day. Jolly's biography of Marconi [Jolly 1972] mentions Maskelyne's collaboration with Hozier for Lloyd's.

[8] The message, "rats", that Maskelyne transmitted carried a cultural meaning. During the Boer War (1899–1902), the British Army heavily bombarded a Boer entrenchment, and asked them, by means of a heliograph, what they felt about the power of British shells. The answer the Boer sent back, by means of heliograph, was "rats". After this, the word came to symbolize "a warning against overwhelming pride". See Austin 1903.

[9] A fifteen-year-old boy with the last name of Bruce had apparently assisted Maskelyne at the time. A half century later (in 1959) he asserted categorically that genuine radio transmissions had been made from the rooftop of the Egyptian theatre. F. Shore (Mr. Bruce's son-in-law) to G.G. Hopkins, 22 April 1959, MCA.

[10] Specifically, though Maskelyne did consciously use an "untuned" transmitter, he probably did not understand just what such a thing would do. His idea seems to have been simply to avoid tuning issues altogether, to use a device that had nothing of "syntony" about it.

[11] The salary of an assistant professor in Fleming's electrical engineering department at UCL was £300 per annum around that time. £300 per annum was exactly the same as what Fleming had been paid while working in the 1880s as scientific advisor to the Edison–Swan Company. The reason why Fleming initially asked such a modest salary in 1899 was that he regarded advisorship to Marconi as a part-time position for which he would not spend too much energy or time [(A copy of) Fleming to Jameson Davis, 2 May 1899, UCL].

SUBJECT INDEX

J.Z. Buchwald (ed.), *Scientific Credibility and Technical Standards in 19th and early 20th Century Germany and Britain*, 177–180.
© 1996 Kluwer Academic Publishers. Printed in Great Britain.

INDEX OF NAMES

J.Z. Buchwald (ed.), Scientific Credibility and Technical Standards in 19th and early 20th
Century Germany and Britain, 181–182.